NEWLY FROM THE SEA

By the same author

SALMON AND SEATROUT IN WILD PLACES
THE ART OF LAKE FISHING
PIKE ON THE PLUG
CLEAR WATER TROUT-FISHING WITH WORM
THE SPENCER LAKE FERTILITY CHART

Frontispiece

PLATE 1

AN EIGHTEEN-POUND SPRING FISH FRESH FROM THE TIDE. A HEN FISH OF THIS CLASS IN THE SHINING BLUE, SILVER AND MAUVE OF PRIME CONDITION IS A DELIGHT TO THE EYE

SIDNEY SPENCER

Newly from the Sea

Fishing for Salmon and Seatrout

H. F. & G. WITHERBY LTD
61 WATLING STREET LONDON EC4

First published 1969

SBN 85493 011 6

Printed in Great Britain
by Ebenezer Baylis and Son Limited
The Trinity Press, Worcester, and London

But salmon and trout play so important a part in the lives of many of us, they are so interwoven with our happiest days, with our hopes, our joys, our fears, our sorrows, that we can never be satisfied.

G. H. NALL
'The Life of the Seatrout'

Foreword

BY BERNARD VENABLES

It seems to me a natural thing that I should be writing a foreword to Sidney Spencer's new book, and perhaps there is a slight oddity in my feeling so. I have never met him. I have never met him, but, for a great many years, among those who have written about fishing his name has suggested to me something personal, almost a tinge of intimacy. I know the origin of it, and the years it covers go back to 1936 – so long ago, and a time of so many changes in things that affect fishers and fishing.

In that year there came a new book; *Pike on the Plug* was its title, and its author Sidney Spencer. It was then I first heard of him, and I read his book avidly and read it several times more. Its effect on me then was profound and remained so for some years thereafter – I was then, I recall, rather addictively fascinated by pike and pike fishing, and the sense, the feel, the atmosphere of the book not only matched my feeling closely, but it introduced me to plugs. The plugs themselves which, immediately, I collected greedily, gave me intense pleasure, so lusciously bizarre were they, but as well they brought me such sport with pike, such a difference of sport, that I was left with gratitude for Sidney Spencer.

But not only that. This was a book strictly technical in intention, and though entirely successful in that, it was pervaded by the feeling of the author's pleasure in his fishing. The atmosphere of it, his zest, his joy in fishing water, came unbidden through the technicalities, a characteristic to be found in the very best of instructional fishing books. Now, so long after, I recall well the sense of that.

It was later – how much later I cannot remember – that I read his book, *Clear Water Trout Fishing With Worm*, and again there was the same lingering atmosphere left in me; and, need I say, in both books the technical purpose of the author was greatly strengthened by this touch upon the imagination.

So now, when so many more years of fishing have passed over him and me, I come to the reading of his new book with an edge of expectancy of pleasure – and indeed of certainty of learning. I suppose too my expectancy has been the greater from conjecturing the effect of deeper maturity and the immensity of further experience the years must have given. Expectancy is not disappointed.

This is a book, mainly, of fishing in the wilder places that salmon and sea trout are to be pursued, and for all who aspire to such fishing, those rocky streams and lochs, often peat-stained, the spate streams, the Devon rivers in their rich plunges of valley, I can recommend no better manual than this. Mr. Spencer has fished so widely, so steadily, so thoughtfully; he has within his record the solving and explanation of so many of the day-to-day problems of individual fish, that the teaching of the book has a penetrative authenticity. And, as I expected, that authenticity is intensified by his retention of atmosphere, his conveying the feeling of each day and place.

I can *feel* the soft scrunch of the heather, smell it and the rivery earth in the summer sun of seatrout time. Maturing has, as I thought it must, deepened those qualities that first engaged me in *Pike on the Plug*. This is a satisfying and lingering book for fishers of seatrout and salmon. I anticipate re-reading when next good fortune takes me to such places as have served him so well.

B.V.

Contents

		page
Foreword by Bernard Venables		7
Introduction		13
CHAPTER		
I	At Some Time in Every Day	17
II	The Reedy Pool	24
III	Choice of Flies	30
IV	Shy Salmon	41
V	Days of the Black Dog	51
VI	Why Does a Salmon Take?	58
VII	'Difficult' Fish	68
VIII	The Fourth of the Month	77
IX	Running Salmon	87
X	Saltwater Salmon, and Seatrout	94
XI	Salmon at Dusk	106
XII	Seatrout in the Night	112
XIII	To Land a Fish	126
XIV	The Tree	136
XV	Tackle for Today	144
XVI	On Strange Water	152
XVII	Lough of the Big Wind	160
XVIII	Boat Handling	167
XIX	Salmon Problems	175
XX	Lough of the Crannog	187
XXI	Injured Salmon	195
XXII	Otters	202
XXIII	The Yellow Stone	208

XXIV	Missing – A Run of Seatrout	216
XXV	What Makes a Good Pool	222
XXVI	Irish Interlude	230
XXVII	Kelts to the Healing Sea	237
XXVIII	'Goodbye Now'	243
Index		247

List of Illustrations

PLATE

1 *Frontispiece* An eighteen-pound Spring fish fresh from the tide

facing page

2 North west wind clouds, hard and glaring 48

3 Salmon seeking their north Irish rivers travel the Atlantic tide-rips below these cliffs 48

4 The gillie uses his own favourite net 64

5 Playing a fish 65

6 Quiet weather means idle boats 144

7 A ten-pounder comes to the net 145

8 A strong west wind, blowing from the narrow lough-head can make this water totally unfishable 160

9 A weir with a simple but effective pass 161

Introduction

Game fishing has changed vastly in recent times. The difficulty of finding fishing and its cost, if and when found, are greater than ever before and there are fewer salmon and probably fewer seatrout than formerly. The erosion of fishing availability must concern us but angling is by its nature a cheerful occupation and these are intended to be happy pages. Here, therefore, we look only at what present conditions mean to us if we propose to catch migratory fish in these islands.

There is an obvious ever-growing need for the angler to make the utmost of such chances as he has. He can best do this by cultivation of *sheer skill in fishing*. There is greater dexterity today in the handling of rods and reels – and also of some gadgetry – than there has ever been, but this is but a little and without doubt the most easily acquired part of what I consider to be the accomplishment of a fisherman. The rest is knowledge of the world of the fish and of the interplay of wind and weather, moving water, of the colours of earth and sky and of light, heat and cold. The more you can think like a fish the closer you are to being a 'cunning fisher'. To be so designated, in my beginning days in Scotland, was a great thing. It meant you had the known ability to catch fish in very difficult conditions, because by some strange alchemy you had the wild waterside in your blood, you had studied your fish in relation to the medium they lived in, you knew not only your flies but the way the fish liked them presented and the inelegance of your casting was no more than an eccentricity which the years would modify.

The dearly-loved sound was the voice of the river and the seethe of waves on some far loch shore, the bubbling of the whaups and the deep croak of the raven. You could fish the evening rise, sleep a rain-filled night in the heather and rise to fish again in the dawn. And you brought home a salmon, or a few seatrout or a basket of brownies caught, because you were a cunning fisher, by knowing precisely how. A man needs to know how today to bring home even more modest abundance of fishes.

This was in my mind when, pondering the form this book should take to be really helpful to those of lesser experience, I asked myself 'To make the most of fishing now, what must a fisherman know?' The answer in any sort of detail, of course, fills the shelves of angling libraries everywhere yet there remain not only great mysteries in the lives and behaviour of migratory fish but such gaps in the craft of catching them that many earnest anglers are indifferently successful. Sophistication in equipment tends in my view to go too far, knowledge and consequent understanding of the fish not far enough. Admittedly, there are not enough years in a lifetime to learn it all and most of us, whatever our experience and cunning, meet situations which defeat us utterly. I am personally all too well aware of some of them.

At this stage in my cogitations and happy in the possession of a wide circle of fishing friends of varying seniority, I began to cast a fly among them. 'Are there,'

I asked, 'any situations which frustrate you entirely, any conditions for which you have no technique or method, any aspects of salmon or seatrout behaviour of special interest?' Some victims remained stolidly in their lies, others swirled half-hearted at the fly and some rose briskly. The brisk risers' problems form the basis for some of the chapters in this book. If I could deal adequately with all that were put to me I should catch a great many more fish than my present skill permits but I shall hope that in trying to clarify some of the more obscure phenomena I shall have made fishing more rewarding for some of my risers and a vast number of others who, were the fly offered, would also rise.

Failure, which stimulates study and analysis has a great long-term virtue of its own. Too many older-time fishing books, in my opinion, exuded repeated, almost constant success and it can be wearisome: fishing in any event, is not all like that, even in the experience of the most proficient of us. Failure is the challenge that will one day lead to celebration deserved but without conceit. A big bag on the day when the fish are 'on' and one could not go wrong is gratifying but no more – far preferable the time when a small or moderate total results solely from application of a lesson derived from previous defeat and directed at difficult fish or poor conditions.

The common belief that the truth is not in anglers is a joke which fishermen bear with equanimity but inside the craft the truth is rather important if we are to learn one from the other. For this reason and because of the usefulness of failure these notes contain, as true records of fishing and the results of experience, something of failure as well as of success.

I have trodden the far-away paths of the otter and I have, for haunting, an unquenchable nostalgia for the wild. If it disturbs these pages I ask no pardon for an endeavour in honesty to preserve something greatly cherished by my kind of fisherman.

SIDNEY SPENCER

Axminster
May, 1969

I

At Some Time in Every Day

WHEN first you fish for salmon, the excitement engendered by the prospect, however slight, of contact with the powerful stranger from the sea totally obscures your ignorance of the diverse ways of the fish. Its comings and goings will be known in very broad outline but its moods, its reactions to the unending variety and combinations of wind direction and strength, water colour and temperature, air temperature, light, and wave-form and height in loughs – all affecting its inclination to take, are in the pages of a closed book. It will take a long time to turn these pages; some are uncut and only the slow knife of experience can part them.

Salmon fishing at its best should remain an excitement, and the element of it to linger longest is in the background of consciousness – the strange, shadowy fear of the fish. Only a salmon fisher can understand this; few will talk about it and fewer still admit to it, but it exists.

In a sense, it is unfortunate that salmon are so unpredictable of mood and that hence so many days show the fish unresponsive and prospects apparently non-existent, because in fact few fishing days are hopeless. A dirty bank-high flood still rising is one thing, a day when the river or loch is fishable but apparently fishless is quite another. The ability to recognize the conditions that should keep one fishing, however fishless the water may appear, is printed on one of the pages staying longest hidden in the book of experience. Even after you have studied it, there will be mistakes and after forty-five years of salmon fishing I make them. It is a question of recognizing in the conditions of the day as they are, or are likely later to be, the factor that justifies hope. We are dealing with a wild creature sensitive to and influenced by forces beyond our perception but we can come to know and detect some of them. Coming to know salmon we learn beyond any doubt that – *at some time in every day there will be an inclination to take.*

I am far from saying that they *will* take but there will assuredly be a time, an interval perhaps very short, in which your chances of inducing a fish to take are at their peak. You should be fishing at that time, not dozing in the boat, wandering off to smell the primroses, or studying the ants creeping in the roots of the grass, but fishing. If you must do these things, they should be done when the hour, for that particular water at that season, is known to be a 'dead' one and when light, water, wind and temperature have been unchanged for some time. Hunger or thirst deprive many an angler of a fish, so lunch at a 'dead' period rather than the regular hour and go by sun time which is nature's time and observed by fish.

The 'dead' time is, of course, any time between the hours which according to observation tend to be the taking times on an average day. The theoretical best taking time, assuming a reasonable fishing day, varies from river to river and lake to lake and in each place with the month of the season. For example in March and

April in Devon when mornings can be cold, it is unusual to move any salmon to the fly before noon by the sun and for the next two hours one should fish hard. As the season progresses and water is 'warm' earlier in the day, one can find fish willing to move at any time in the morning and there will be another but less marked move about three-thirty to four p.m. and again at sunset. Where water temperature is really high in summer and early autumn and water low, there can be no 'good' time except perhaps just at or immediately after sunset. In such conditions other factors operate – the variables in wind and weather, air and water temperature, which can blow all the 'dead' time and 'good' time ideas to ribbons.

But, still looking at this question of time in isolation, I turn from my West of England example to that of a western Irish water. Here in March and April you will be fortunate to raise a fish to the fly unless, towards the end of April, warmer than normal weather prompts a slight move in the day at about three-thirty sun time. May/June provides two 'good' times, the first about eleven-thirty and the other at four p.m. In summer one can find a taking fish at any time up to about six-thirty p.m. but rarely later. There is no sunset take whatever. Strangely the seatrout, which are numerous, also go off at about six-thirty and there is no evening fishing. This is true of seatrout also in some rivers and lochs of the Hebrides though here salmon will come to the fly in the evening.

So much for time of day. By itself it has little significance and I would rather be on the water in a reputed 'dead' time on a really good fishing day than at a 'good' time on a really unfavourable one. Time of day is no more than one factor in a whole array of them.

The others I have already touched on – sky cover and colour, water and air temperatures, water height and clarity, wind direction and strength. A wide range of combinations of these will constitute favourable conditions, that is to say, will predispose fish to take, but certain of them, notably wind direction, strength and temperature, of course may vary in effect with locality and topography. A wind that will raise a nice ripple against the current in some alignments of river pools may blow one off the loch or, offshore on the best fishing ground, may not give enough ripple. Easterly winds are, generally speaking, bad at all seasons and northerlies not much better in spring, but sometimes not by any means so bad in summer. North-westerly winds, in most parts of the British Isles tend to be hard and gusty and while this can be no more than a nuisance to casting on a river, it can ruin prospects on the loch. A wind which keeps swinging a point or two and back produces on the lake a little wave on top of the true wave and this is bad. Even worse is the black gust that skitters here and there and while it passes over the boat and nearby water, makes it impossible to fish the fly properly.

Southerly or westerly wind sources are demonstrably best and of the two I prefer the west and there are not many places where fish, other factors being equal, do not respond to a steady west wind. Again, it is of course, possible that local land contours may blank off the west. I know a part of one lough where a strong west wind, kept off the water by high ground, strikes in from the north round the shoulder of the hill. One can multiply examples indefinitely.

The effects of wind direction on temperature, initially

of air and in course of time of water, are least marked in locations far inland since they reach the water over appreciable land surfaces. On coastal rivers and lochs it seems to me that they tend to retain the character inherent in their source. In summer in the Hebrides I have known south winds bitingly cold after making a long ocean sweep and warm, almost muggy easterlies reaching the islands across the warm Scottish mainland.

The winds bring the cloud and the nature and density of cloud affect the light. East wind frequently breeds all-over cloud cover giving grey days and in late summer, haze, and sometimes enduring for a long time. The north can produce anything but not often complete cover; clouds tend to be hard of outline. Hardest clouds in towering castles of white, blue-black squalls and fluctuating temperatures characterize the north-west blow, though really hard-edged cloud masses from any quarter tend to produce a glare which keeps fish down.

The light that it commonly produces almost everywhere is one of my own reasons for preferring the west wind above all others. It is a soft light and cloud-masses are soft and broken so that the light reaching into the water is diffused, bluish, broken and ever-changing. It is the constant change that provides the stimulus to move in the salmon. I hope that a beginner reading these notes will not conclude it useless to fish other than in a west wind – I am making fine distinctions but there are fish to be caught in winds from any quarter – even the execrable east because, of course, on any given day, factors other than time of day, wind direction and light, which we have just looked at, may be very favourable.

Your river or loch may, for example, be full of fresh fish with water falling and clearing after the spate that brought the fish. I would fish in any weather conditions in such a case. I would fish and I do fish on many days of desperately unsuitable winds: I merely say that if I have a choice I like the west wind.

What of water height and colour and clarity? So far as rivers are concerned I have talked about such generalities as are useful in the chapter on 'Running Salmon'. Optimum heights of water vary tremendously from river to river and are in some degree related to clarity. Dependent on the source and drainage area of the river and the nature of its course is the quantity of sediment or stain it carries and the rate at which in the first case, it will settle out and clear, and in the second at which it will flush away. The most reliable guide for the stranger is usually local information which often consists of dicta like 'When the white stone by the third pier of the bridge is just showing, she'll fish!'

It is somewhat easier to give guidance regarding height of water in lochs. For one thing it is less critical. There is always a level at which a loch tends to fish best – or at which the usual fishing grounds do best, but there is a wide tolerance and a loch somewhere round about the reputed best fishing height is in order. A very high loch seldom fishes well and I would rather have the water on the low side once fish have come in from the sea. There is some variation, of course, and I know one loch in the far north that is utterly unresponsive when filled up by a spate and wonderfully productive when low level rocks emerge from shrunken waters and another in the west where I have taken many salmon off flooded beaches. But generally, medium level to low water is best.

Colour in lochs is seldom more than a fleeting

problem. The area about the point of entry of a river stained with mud or peat washings can be dirty and filled with the usual debris of floating dead grass and heather and the flotsam can be a nuisance as far out as the river current persists on the surface. But a lough is a vast settling tank and dirt in suspension soon settles. Peat washings are another matter, the dark water they produce may hang about for days. This sour peat water affects the loch as it does the river, putting fish 'off' until it clears. If there is any colour at all, whether from sediment-laden field washings or from unsedimented peat water I like to fish either Fiery Brown or the old Irish Claret somewhere on the cast. Both were devised for this purpose.

The bulk of fishing lore, much of it unwritten, is immense and the greater part of it concerns weather. In so far as it embodies sound conclusions derived from long observation of factors which unquestionably affect fish behaviour it deserves study. But the longer one fishes the less dogmatic one is inclined to be and the more likely to indicate simple guide lines as I have tried to do. The ifs and buts, exceptions and combinations are unlimited, the pattern is elusive and only much fishing, good observation and a retentive memory make sense of it.

Perhaps it is best to say that in the long process of arriving at knowledge of salmon ways, one learns beyond any doubt that at some time in every day there will be an inclination to take. Its intensity is governed by conditions as defined above.

II

The Reedy Pool

IN A mountain river that I know, there are two very big pools, two spreadings of the river extensive enough to qualify as small lochs. A hundred yards apart, they lie in hummocky moorland three miles from the sea. They are great holding-places and always contain large numbers of salmon – more often than not exceedingly difficult to catch. I know of only one man who caught them 'on' and he had seven in three hours.

The upper pool, resting place of vast numbers of fish, is round and open and because of its openness to every breeze that blows normally carries a good ripple and often a really good fishing wave. Add to this the fact that fish prefer it to the lower pool – possibly because of its cleaner bottom, and it will be clear that generally it is considered the better pool of the two. It can be frightfully dour however, even when well-stocked and in good fishing condition. It has, as a fact, the lesser part in this account of a day but if there is to be any point in my record of what happened, you should be able to imagine both pools.

They could scarcely differ more than they do in all their features including the behaviour of fish lying in them.

The lower pool is pear-shaped with a shallow bulbous outline in its lower half, and a narrow upper part tapering to the point of entry of the river. The bulk of the pool is weedy and reedy – reeds in forests over much of the area and scattered over the rest with, in places, dense filmy underwater growth as a further barrier to the working of a fly. The narrow upper part is deeper – inky-black peat water. Along one shore and growing on a submerged peat ledge is a dense jungle of reeds with a sharply defined edge where the shelf ends over black depths. Because of the reed barrier this shore is unfishable but the opposite bank is stony under a steep bluff and can be fished. Downstream the neck of the pear gradually widens to the shoulders where the entire breadth of the pool is closed by a reed barrier. The fishable area is therefore small. There is no pull on the stream at all save just where the river enters, the only evidence of current elsewhere being the all but imperceptible tremor of the reed stems.

But the greatest superficial defect of all from a fishing point of view is the blanketing effect of the bluff already mentioned. Nine days in ten there is not the ghost of a ripple except perhaps in the rather more open but quite unfishable lower portion of the pool. Fish show continually in relatively small numbers, mostly cruising or splashing along the edge of the reeds and just about beyond reach from the shore below the bluff. Very few fish are taken from the Reedy Pool. Its apparent difficulties deter most rods and few give it more than a desultory cast or two in the short current at the head. The upper pool is, after all, the great attraction since, unless it has been overfished, it seldom fails to yield at least one fish the first time over. And fish show there all the time so long as the boat is not permitted to cross the main lies.

What usually happens is that one gets a fish or perhaps two immediately on arrival at the upper pool which then goes 'dead' for the rest of what has to be a very short fishing day because of the very long mountain walk involved to reach these pools.

For years, like everyone else, I used to fish the upper pool until the fish went down and 'off' and then turn homewards, stopping only for a few casts in the fast-running neck of the Reedy Pool on my way downstream, and dismissing the rest as hopeless. After all, many whose skill I respected had spent hours vainly trying to lure a fish from those dark still deeps. There came a September day, however, when the best I could extract from the upper pool in fairly good conditions, was one small fish and with an hour of fishing time in hand I paused a little longer on the bluff to look down on Reedy.

Five fish were showing – two fresh and three very dark ones. They played follow-my-leader round the open water, appearing first where the reeds shut off the lower part of the pool, cruising slowly upstream and head and tailing along the ledge where the reeds ended. Then would come a splash by the stones where the river entered and a slow cruise down the stony unreeded shore to the end of open water again. The fresh fish were slightly the more active and showed now and then off the end of a rock rib which jutted diagonally into the lower part.

Where, clear of the bluff, the wind struck the far end of the pool there was a faint ripple among the reeds but elsewhere not so much as a crinkle stirred the surface. 'If only we had a breeze!' – I said to my gillie – 'those head-and-tailers should take.' It was comfortable there in the dry heather and for as long as it takes to smoke a cigarette I debated lazily whether I should make up a lighter cast with very small flies. This seemed the only hope but sheer indolence won the day and I was presently crouched on the rock rib and waiting for one of the fish to break water above me and give some indication of the path he would follow on the downstream leg of his circuit. He porpoised five yards upstream and I put the flies, a Sweep on the tail and Butcher on the dropper, both 8's, across his line, working them smoothly and slowly. Either he did not see them or was not interested because he showed again far below me. I left some line hanging loose in order to show him the flies when he turned upstream again and this time they were a yard in front as he head-and-tailed opposite me. Again no result but recovering line very slowly with the flies coming in to my feet, I was taken solidly by quite a different fish and one I had not seen.

I thought he would dive for the cavern under the peat ledge along the other bank, but he did not and we presently netted a fresh eight-pounder with the Sweep in the corner of the jaw. The others had gone down and I gave them ten minutes before starting again. Again the same procedure, casting slightly upstream and across towards the reeds – the fly travelling very very slowly and very deep on the retrieve. The second cast produced another fish also fresh, and a crazy one. Indeed he behaved more like a seatrout with leap after leap, the last of them a lightning-fast cartwheel away from me on a shortish line. I was slow to ease the pressure and he broke me. That finished the day. And all the time the pool was dead without current or a crinkle of wave.

I was more interested as was my gillie to get fish at

all from Reedy than annoyed to lose the second one and anyhow one fish from each pool was a great deal better than a blank day. Perhaps we had not until now known the taking place in Reedy Pool. I would put it to the test next day.

And I did. The day was different – bright where it had been dull and with even less wind and again I had one fish, before a late lunch, from the upper pool. This fish took a Dusty Miller which suited the sun and the crisp ripple and I kept it on, with a Torrish on the tail to fish the Reedy Pool. But the Torrish looked garish over the dark deeper water and I replaced it with a Black Doctor. Again the flies were 8's in spite of the calm water. The cruisers were active as before but seemed to follow a narrower ellipse far out so that none came near my rock rib. Moreover all the fish showing were stale and dark. I began with little hope, throwing a long line to the cruisers along the ledge and fishing diagonally upstream and across. I was thinking after ten minutes or so, that yesterday's experience had been a fluke when a very dark fish broke surface within easy reach and with a slow wave of a black tail was gone: a hastily shortened line put the flies in front of his last known position and without marking the surface he was on.

This one led us a pretty dance. He persisted for some odd piscine reason in running into the shallows, returning time after time whatever I did by way of side strain to keep him out, and was finally got over a sunken net whilst swimming strongly in water barely over his back. Nine pounds and on the Dusty, and as black as a fish can get after a month or two in that kind of peaty water.

Some sort of pattern was beginning to show itself

though I was puzzled by the fishes' willingness to take what, on the face of it, was much too big a fly. Nor were they put off by the weight of cast necessary to carry the fly. The simple requisite was that the fly should be deeply sunk – I reckoned mine were nine to twelve inches deep when taken – and proportionately slowly moved. The fact that the flies in all cases were fished across the path of the fish is of doubtful significance. Because of the reeds and the cruising line of the fish I could only present the flies this way, though admittedly the angle of draw that this involved gave the fish the best view of the fly.

But anyhow, here *was* a pattern – and a method that would extract salmon from Reedy Pool and I kept applying it. Ten minutes' rest after playing a fish and then when they had settled again, very quiet casts cautiously made from concealment until a willing taker was intercepted. On the day I am describing – my second attack on the pool – I had two more fish and left the scene with three fish and a great conceit of myself! The heavy fall that should have followed my preening did not materialize either because on subsequent days I continued to take fish until I had had nine and lost two. Conditions throughout were by accepted standards bad. It had proved impossible to take more than the odd fish from the upper pool in similar conditions in spite of its much larger stock and better reputation for taking, but for my part I am very content in the Reedy Pool whenever it is my good fortune to fish it.

III

Choice of Flies

ONE of the discoveries which every keen fisherman for salmon and seatrout makes as a long succession of seasons passes by is how few patterns of flies are really necessary for the waters he fishes. Naturally, if his is fortunate enough to be able to range far and wide on rivers and lakes from the far north of Scotland to the Islands, to Donegal and Connemara and Kerry and back to Wales and the West Country, he will need many fly boxes to house his bare requirements. But for each individual fishing the variety of flies will be small, and since most men have favourite waters or places where they enjoy some rights or privileges of fishing, the accumulation of vast ranges of patterns does not arise of necessity. It may and usually does arise from other causes, and every devoted angler has scores of flies that he never or very seldom uses – flies bought on someone's recommendation – on the strength of one good day with some special pattern – or in conformity with local fancy. It is an expensive and rather silly practice and I am as foolish as anyone. Every so often I have a clearing-out but I lack the courage (or the heart) to discard from hundreds of flies I shall never use, beautifully tied, new-conditioned dressings bought at some time as being absolutely 'the' fly for such and such a water. The trouble is that I did not buy three and try the fly – I bought a dozen and still have eleven of them – bright, attractive in their way, and useless.

There is another difficulty in the way of being hard-hearted and sensible in keeping the numbers of patterns down and that is the fact that 'odd' patterns do sometimes kill fish. Whether they are established local dressings or just someone's experiments does not matter and because they once served well, they will be retained with a certain affection. I wrote at some length on the subject of limitation of the number of fly patterns in *Salmon and Seatrout in Wild Places* and I suggested a short range of flies for specific conditions of weather, light and water colour. I do not propose to cover that ground again but I will traverse an adjoining area which will include a slightly widened special-conditions range and a few local patterns which have proved to have wider use. In so doing I do not retreat from my view that most of us carry far too many patterns and that variety should reside in size range rather than in dressings. After all, all the dressings that ever were will not do you any good if they are all on size four or six hooks and you are on a little shrunken river in August sunshine, but you have some hope if you can go down to nines in any of half a dozen patterns or even fewer. I can hear someone say, 'That's all very well but who would be so foolish as to go out so ill-equipped for summer fishing?' and my reply has to be that I have seen it done, out of miscalculation of conditions on a far-away water or lack of local information or good sense or all of them.

There is little point in an attempt to define a short standard range of patterns for general use. To do so with any usefulness to the individual one has to know where the bulk of his fishing is done. The fly that will take salmon anywhere *may* exist and if it does will be found in my opinion among Yellow Torrish, Thunder and Lightning, Dusty Miller, Black Doctor and Silver Grey. A range, in assorted sizes, for the man who fishes usually in one river or lough somewhere in these islands might be these flies plus local patterns – or other standard dressings which kill well locally because they suit water colour or for some undissectable reason. Perhaps – and it is a highly gratifying phenomenon – the fish just like them.

Thus acknowledging the problem presented by the all-purpose, all-locality standard range, I must say that I have found a few flies devised to meet purely local 'tastes' which seem to have some general usefulness within my own experience. That there are many of these which are seatrout patterns is odd, because the laying-down of a standard seatrout range is a much simpler matter than is the case with salmon flies – the basic dressings usually being acceptable to seatrout in spite of the multiplicity of variations and purely local fancies. A fly box containing Butcher, Mallard and Claret, Woodcock or Mallard and Yellow, Blue and Silver, Zulu, Blue Zulu, Connemara Black and Peter Ross in assorted sizes would equip me to fish wet fly confidently anywhere. On the basis of sound and tested local information I should probably add the river or loch special particularly if I were night fishing.

This business of fly dressings, hook sizes and shapes is probably as productive of amiable wrangling among fishermen as any of the fishing subjects which abound in ifs and buts, exceptions and reservations. Someone will pick a fly from someone else's fly box and say, 'What do you call that?' 'That? oh, that's a Green Imp – Brora fly you know. Funny thing, they won't have it without the guineafowl. I remember . . .' And then it starts. Another man says, 'Down on the So-and-so in Kerry we always use a Coiner with a dark hackle. I was there one day and the river was a bit coloured . . .' More fly boxes are produced and more tales of super critical fish are told. Ours is a happy pastime.

I must go back now to the question of suitable patterns of salmon flies for special conditions, to be added to the standard range of your fancy and your need. To Lemon Grey for misty weather and filtered light, Dusty Miller (primarily as a dropper in lough fishing) for bright weather and Fiery Brown for high peat-stained flood water I will add two for dark overcast days and very rough water and two for similar conditions but quieter water. The first pair consist of Black Fairy, an Irish pattern of general usefulness on days as sombre as the dressing – which is for all practical purposes a Mallard and Black with black hackle, and another Irishman, the Claret. All that need be added in respect of the Black Fairy is that the wool body should be roughly teased out between the turns of gold tinsel. The Claret, not often seen nowadays, is dressed with dark mallard wings and a few long fibres of jay. The body is made up of a sparsely wound burgundy-coloured hackle overlaid with four or five turns of flat gold tinsel. There can be, but need not be, a butt of black herl and the tail is a few golden whisks. Hackle is jay mixed with the first few turns of the burgundy body hackle. This fly, providing the body winding has the true prismatic quality, glows like a dark jewel even in bad light. It will, like Fiery Brown, do well in peat-stained water. The other pair, doing better as already noted in quieter water, possibly because of their better entry are Thunder and Lightning and Sweep. Both are standard patterns and require no description and both are bad light flies. Sweep, indeed, often fishes well in darkness if you have a mind to try for a salmon so late in the day. As an after-sunset fly when salmon may be moving but the light has not yet quite gone, it can be excellent. I have done well with it in the throats of river pools when the fish have been taking their evening cruise, and in the current where a river enters the lough. It seems to do best in sizes not larger than six. In reasonably broken but not turbulent water I should fish an eight. I think perhaps that to my dark weather list I should add Black Doctor, a fly most effective in the Hebrides so long as it is fished in small sizes. I have had many salmon on sizes as small as ten but I have only once taken a fish on a size over eight and that was in the West of England. I remember that fish very clearly. Tree-cutting had been neglected for a few years and one bank of the beat I fished had become overgrown. Cutting was in progress when we had a big spate and work was stopped with one tree, an ash which leant out over the water, left uncut. Some way upstream, in the cleared stretch, fishing a 1/0 Black Doctor I hooked a big rough-fighting fish which could not be persuaded to go up nor prevented from going down with all the push of a swollen river behind him. I had to follow and in due course arrived at the tree. I could just keep the line clear of the outward-leaning trunk but I had no hope of passing the rod round. With all my line and most of the backing gone I chose to hold on and after some cavortings which I could feel but not see, the fish stopped! I then very gently walked him back until I could afford to use the reel and so got him. The incident did not however convert me to the use of big Black Doctors. I think it was a question of size and of visibility in rather thick water and that any other really big fly would have been taken by that fish.

Golden Olive, again an Irish pattern, has been serving me rather well in recent years for both salmon and seatrout – the salmon mainly in Ireland and on loughs, and the seatrout pretty generally. The yellow wool body shows up well and the fly almost always *looks* right in a variety of lights. In one lough I know where salmon show a decided preference for light-coloured flies predominantly yellow in tone I would rate Golden Olive and Yellow Torrish as equal best. But, fishing both on the cast as I often do, I like the Olive, which has more weight when wet, to be the uppermost of the two because a bulky-bodied dropper tripping rather better brings up more fish than a clean-lined fly creating less surface distortion. It may be a fine point but fishing is full of them.

The Olive can readily replace Mallard and Yellow in the standard seatrout range as the differences are small. My own choice is the Olive which I carry in small sizes for summer salmon anyhow, and therefore have the fly in sizes suitable for seatrout whenever I wish to include it on a cast. While we are in the homeland of the golden fly I ought to list Connemara Black which seems to be more widely known and appreciated than formerly. Thirty years ago it was hardly seen outside the west of Ireland but now it is often seen in Scotland. There are few really dark seatrout flies – one thinks off hand only of Butcher, Zulu and Black Pennell, and Connemara

Black fills a gap in that among dark flies it simulates the natural fly more closely than others. As an imitation of a number of dull-coloured insects which may find their way on to the water it will not, of course, bear close examination. It is a good and reliable pattern everywhere that I have fished it, though for river work I like it rather more sparsely dressed than is usual.

You may not, on this side of the Irish Sea, find all these flies in stock in any and every tackle shop but they are regular and recognized patterns and are obtainable.

There are a few more that are specials but are worth having, especially for night fishing for seatrout. The best of these is a north country fly widely used for trout and deadly for seatrout in real darkness. It imitates if anything, the larger sedges and, apart from the wings the dressing does not seem to be critical. The fly, Brown Owl, has wings from Tawny Owl lesser coverts, orange-yellow silk body and a rather fluffy hackle of soft brown like the wings. Some I have seen, by amateur dressers, have a yellow floss tag but there is a variety of ideas about this fly. As a rule the result is an untidy, shapeless creation which however spreads in the water and gains silhouette if not shapeliness. It is used for trout in small sizes in early summer but seatrout take it well at night up to the beginning of September.

The fly called 'Queen of the Water' or 'Queen of the River' is consistently successful with me for night fishing. I think it originated in the West of England where one still sees it most commonly. One version has clipped wings and is too stiff as a result, in my opinion, but the dressing varies very little. Wings are of very pale mallard with faint flecks on a creamy-grey ground, and body all over flat silver tinsel widely overwound with round silver wire or fine cord. Hackle is honey-coloured. I have found it fishes best unworked, drifting and not larger than nine.

Very pale mallard is also used for a fly which to me is nameless. It is a simple variation on Mallard and Claret in that the wings are, of course, much lighter in colour and the body is very dark claret wool not picked out between the four or five lines of broad silver tinsel, and therefore compact. Hackle is black and tail two or three fibres of tippet. Again it is a night fly and I always use it in a fairly big size, say seven or eight, when I have reason to think there are big seatrout about, since some years ago I seemed to detect that it had an attraction for the bigger class of fish. Every fisherman has his fancies and perhaps that is just one of mine but it is certainly true that I have had many big seatrout, fish up to nine pounds, on this fly.

Whether seatrout are more discriminating in their acceptance of flies than they were say, seventy or eighty years ago, is doubtful but my own view is that salmon certainly are. I think that I can see a great difference in the forty-odd years that I have been salmon fishing because it is quite certain, for one thing, that the size in *general* use then, will not take fish today except in heavy water. And if you go back to the turn of the century or earlier, in the illustrated books of the sport you find depicted there flies of sizes and shapes that look ludicrous today. Materials were less exotic, there was a predominance of turkey as a base for wings and perhaps this was a good thing in one respect because at a later stage and indeed until recently, the bulk of tied flies were quite unnecessarily elaborate. But the very early flies, as used two generations ago, would not kill salmon today other than in exceptional conditions. They do not look right – just a little ridiculous. Therefore one assumes that the 'taste' of the fish has changed and that their discrimination is greater. I find it difficult, however strong my conviction that the fish *have* changed, to accept sophistication as a reaction of short-lived migratory fish of double and seasonally changing environments.

It is a fact also that in all fishing the fly that is most fished will over a period take most fish, so that if there are angler-made fashions in fly patterns and fly size, there will be enough success with them to make it appear that the salmon have changed their 'taste'. Have flies changed over the years because the older types gradually ceased to catch fish, or have they changed at the whim of the angler?

I suggest that in the heyday of salmon fishing, when the whole conception of the sport had an element of fashionable conduct and convention-dictated technique about it, the fish were so much more plentiful than they are today, that gear and flies which we would call downright unsuitable did catch fish. There is besides, some evidence that the spring fishing catches in those days as recorded included many kelts. In short, I suggest that stocks are so much smaller now that the uncritical and therefore easily-catchable element is very small indeed: bit by bit the ways of tempting the 'discriminating' balance have inevitably been refined so that the great clumsy flies of yesterday have gone, along with the heavy casts, lines and rods, to an infinite increase in the pleasure of fishing.

Anyone who commits himself to paper on the subject of salmon behaviour exposes himself to the hazard of contradiction by the fish themselves. I fancy that if one fished with a wandering microphone attached to one's person, speaking one's deliberations about salmon as one fished, the contradictions would come within five minutes of the statement. Some aspects of salmon behaviour are constant and predictable, others are neither. The 'others' are only too familiar in effect – their causes are the subject only of speculation.

In the present case, having inveighed against our ancestors' crude lures I am bound in honesty to say that I have known salmon to take things just as absurd in appearance as some of the old flies. Two instances come at once to mind. The first concerns a day on a Scottish loch when conditions were about as bad as could be. Wind was weak and light much too bright. Apart from that the fishing ground was deep – and fish will not come to a fly in deep water unless they are 'up', which they seldom are in bright light and near calm. I decided, when most of the day had been blank, to try casting on the move and, moreover, to try something outrageous in flies. I put on a heavy cast which looked like a ship's cable in the water and a huge Alexandra on a 1/0 iron. Its last victim had been a seven-pound pike incidentally. The gillie rowed haphazard at moderate speed while I cast this contraption at right angles to our course and in a few minutes I had a violent take from a twelve-pound fish which was duly boated. This on a day which in human judgment was one for fine gear and some restraint in fly size. My guess is that swinging movement and large size brought that fish up. We did not get another.

In the second case, the gillie, working with me for the first time, plainly thought I was crazy, when I fished a bright yellow monstrosity just for the hell of it. Place was the Hebrides and a very good loch under a fair breeze but rather bright light. I had expected fish to move to the small flies we normally fished because the wave was good enough if not ideal. But they did not – for one of their unfathomable reasons. As a rather idle experiment to see if the fish were 'down' because of the light I unearthed a vast fly from the 'odds and ends' fly box, and put on a heavier cast to take it. It had bright yellow 'wings' of hair, a silver body and a blue hackle on the longest shanked fine wire hook I have ever seen. I cast this among sunken boulders and over a recognized holding area and presently had a somewhat red cock fish lift up from the shallows and rush at what I later called the Yellow Peril, with open jaws. We netted him after the kind of fight these cock fish usually give and in which the heavy cast enabled me to be rough with him.

These two fish were clearly in the wrong generation – or were they?

IV

Shy Salmon

SALMON are strange creatures, strange in a hundred ways, but here in this particular note I have one aspect of their oddity in view – their shyness in some situations and circumstances and their indifference in others. Seatrout do not exhibit the same contradictory reactions since they are consistently shy and indeed, in my view, the shyest fish that swims with the possible exception of heavily-fished and therefore very sophisticated brown trout. Not all browns are, however, by nature as shy as all seatrout. Of seatrout more anon.

There is one milieu in which salmon vie with seatrout in shyness – the very small river. Such water generally lacks depth and cover and often in bare rocky rivers the two are one. Very small rivers in my context are not the tributaries of large rivers – the headwaters where fish can be found from September to season's end – but the little rivers which are truly small throughout their length. There are scores of hill rivers, along the western coasts of these islands especially, where the point of entry to salt water is not more than, say, thirty feet wide. Most are less.

If, in fishing such little rivers, you are so proud of your status as a human animal walking upright on two legs that you cannot revert to all-fours progress or even proceed on your stomach with elbows as a means of locomotion, you had better stay on the big river or the loch, leaving the small pools to the man who is prepared to stalk the salmon and lose himself in his background. I have known a narrow rocky gut containing five salmon and a number of seatrout to be reported as empty by an angler who walked up to it to stand like an animated fish-scarer on its brink. 'Fine and far off' is not quite the slogan but something like it. To fish a short line can be awkward but there is no reason why some line cannot lie across grass or heather so long as the fine end and the cast fish properly and are picked up before line is so short that the flies catch up on the edge of the bank. Freeing this can be awkward if you are to remain hidden. To give you any chance at all the fish must remain in their lies, some of which will be taking places, and if they are to do that you must creep and crawl and lurk behind heather and rock and always be, rod included, below the skyline. There is one and only one type of small pool which can be approached fairly closely and that is the one where the rush of water from above is strong enough to agitate the entire pool surface so that it is entirely broken, foamed and bubbling. I have sometimes taken fish in such cases while standing over the lies.

On open pools, however, the need for concealment cannot be over-emphasized. A problem can arise where the banks fall steeply to the water's edge and here one can only hide against the background. In such places white shirts and collars stick out like sore thumbs and every item of clothing should be in soft browns and greens – I do not like black oilskins either. Ladies, whatever their relationship to the angler, should be sternly disciplined in the matter of clothing colour should they choose to come along 'for a picnic'! Obviously the smaller the water the nearer the lies are to the bank but clothing colour is a serious consideration on all waters – big rivers and lochs not excepted.

I am not sure that I have done the small river problem justice. Because lies are limited by pool size and character and the general lack of depth, fish will be in places which are not recognizable as lies. A stone a foot in diameter will make shelter for a fish in a very thin shallow. Little rivers usually, but not always, carry a small local race of salmon and these fish of six or seven pounds can hide in very shallow water so long as it is rapid-flowing. Isolated lies of this sort take long observation to identify and are difficult to fish properly – the fly only fishing for a couple of feet perhaps before being whisked away by the stream. I can remember taking salmon from a certain long gravelly stickle in a West Highland river. The pools immediately above and below are well-known but I do not think a salmon had ever been killed between them until I knelt far back from the stickle head one day and let a fly work among the stones. Twenty yards of this – very hard on the knees! – produced two fish played without standing up and kept quiet by a light strain and netted by the gillie lying flat with the net on the stream bed. On that day the shallow was not more than a foot deep at its deepest and then only where current turbulence at the rear of a stone had made a slight hollow.

It is quite amazing where fish will lie in small water; they can be taken if you keep away from them. If you want to see where and how shy they are – show yourself and watch the shadowy shapes flicker away. Then go and fish somewhere else.

Salmon on full-size rivers are less easily disturbed. They are less likely to be forced to lie in shallow open water; they usually have more water over them and the angler, however brash, is operating farther from them. Nevertheless it is downright foolish to expose oneself on the skyline by a pool, to wear conspicuous clothing or to tramp heavily and crunchily over banks or gravel. Stealth may not be the critical element it is on the little river but it helps. The less *any* pool is disturbed the better it will be. Mention of gravel reminds me of wading and the damage that can do. I never wade unless absolutely necessary, indeed I regard waders mainly as weather protection in very wet conditions, to enable one to move about dryshod in inches-deep margins and only to be used for their designed purpose where it is impossible to show the fish fly or bait by other means. It is too easy to disturb fish by wading into or near the lies and the difficulty is that the relation of the lie to the bank will vary with water height. A change of lies is dictated by the configuration of the banks and bottom and the effect of the weight of current on them – sometimes a rise of a few inches will vastly multiply the push of water so that fish retreat to the thinner margins. Here, wading in such conditions, you can easily and unknowingly wade among them. Not so long ago I did this very thing. In an endeavour to reach a very awkwardly placed taking area a very long cast downstream, I pushed through the only gap in bush-lined banks into eighteen inches of rather weedy water to step right on top of two salmon lying close to the bank. Admittedly they were in an unfishable situation and I could not have shown them a fly had I known they were there, but it could just as easily have happened in open water. That day, with poor light and a cold north east wind, incidentally held two morals, because I fished hard until four p.m. when I joined my wife in the hut for tea – having decided to call it a day. After tea, however, I was persuaded to give it another quarter of an hour at the tail of a nearby pool while the tea things were being packed and was back at the hut inside the time with a seventeen-pounder.

From shy salmon in rivers we should turn to lakes, lochs or loughs and in so far as these may be fished from the bank one must study to be inconspicuous and quiet. In practice, of course, little bank fishing for salmon and seatrout in lochs really exists on a worth-while scale, though I confess that when I see a boat fishing the shallow shores or do so myself from the boat, I sometimes wonder whether results might not be better from the shore. But I would rather fish on lakes from deep into shallow and best of all *along* the critical edge of the sub-bank some way out from the true shore. Isolated experiences do not prove very much, but for what it is worth I can recall five salmon taken in three days by a bank fisher from an Irish lough shore also fished, but quite without result, by four boats. But, and it's rather a big one, all five came from the same group of rocks lying a comfortable casting distance offshore and much easier to fish because of local circumstance from shore than boat.

Sometimes, especially on small loughs, a lie may be so close to shore and so placed as to be dominated by high banks and rocks, a peat hag or bluff. Such places are to be avoided. Water is generally clearer than you think and skyline movement will put fish 'off' particularly if they are near the surface, as may be evidenced by cruising. Where lies are far offshore, of course, you can walk the skyline as much as you like and your caution in the approach is expressed in terms of boat-handling.

There is something, not much, that you can do to make a boat less conspicuous. It is rather a large piece of wood, or fibre-glass or perhaps rubber – a bulky opaque object which drifts or is propelled across the salmon's window. In my view, after fifty years of boat-fishing, the greatest problem is glare. White or brightly-coloured boats are bad, as are those, for a different reason, of black or of very dark colours. Best of all I consider to be soft darkish grey. The light does not glare on them and they blend with water and sky colours most unobtrusively. I am speaking of bottom and side outer planking of course – the inside can be striped like a zebra if you so wish. Lastly on boat colour, do not destroy the effect of your grey camouflage with a tasteful stripe of bright colour along the gunwale! And if you ever find yourself the possessor of a pair of plain varnished yellow oars – paint them grey at once. If you think I exaggerate the importance of colour in this context – make a simple experiment by discarding (for the time being) the water-bleached indeterminately hued cordage of your landing net. Put on in its place one of the bright orange-coloured nets they seem to sell nowadays. Fish, catch and play your salmon and bring him to the net and there watch his reaction when he sees it. A tired fish that will come quietly over the old dirty-string net will, unless he is utterly finished, go frantic at sight of the bright one.

I do not think that the quiet passage of a boat disturbs the fish to any extent but careless oar-handling and banging about in the boat will, and the man who fishes standing in a boat has least chance of all for he is the skyline arm-waver to outdo all others. What I said earlier about inconspicuous clothing has application in boat-fishing too.

It is in boat-fishing that some of the contradictions in salmon behaviour become most apparent in my eyes because whilst I will not retract one word of my strictures on the general need for quietness or circumspection, I am bound to admit that there are times when salmon will show the utmost indifference to externals. In my own experience uncountable fish have risen with decision and taken the fly inside an oar's length from the boat. In bright sun and dead calm water, with the boat motionless and both angler and gillie half asleep in the hot sun, I have brought a fish up to a slow deep fly and had it swim on the surface as the fly came up to the boat. It took and was hooked with the rod as far back as it would go. That fish was not alarmed either by the boat or our movements in it.

Perhaps the whimsical oddnesses of such salmon behaviour fluctuate with the intensity of the taking mood. The fish that takes in a bright flat calm is 'on' – the fish that flees from a passing shadow and will not take when there is nothing in light, wave, or temperature to prevent it, is 'off'. But since the only positive indication of a take is the appearance of a fish below or behind the fly it must be, on balance, wisest to cover the unseen transitional stage by assuming all salmon one approaches to be very shy.

I once saw salmon scattered through a long pool of medium depth prodded into their refuge – its deepest downstream area, and the shallower-lying fish certainly fled there in disorder. But within ten minutes two of them had taken the fly. To begin with these fish were asleep and, absurd as it may seem, had every appearance of being 'angered' by the disturbance, and then when the fish were herded together in the relatively small area of deeper water, the competitive element played its part in the reaction to the passing fly. Commoner aspects of disturbance reaction are well known to loch fishers. The passage of a drifting boat, particularly in a light breeze, will often result in a fish plunging behind the boat after it has passed. The boat crew looks round hearing the splash and says, 'We must have passed right over that one.' Once and once only, I hooked such a fish by means of a frantic back-handed and very awkward cast which must have put the fly close to the fish before he went down. Usually fish disturbed in this manner by the boat are not takers.

I have never known seatrout to exhibit any of these inconsistencies though there are plenty in other aspects of their behaviour. They are consistently shy whether in the mood to take or not and everything that has been advised in avoiding disturbance of shy salmon applies with greater emphasis to seatrout. An ill-considered and careless movement by a pool – by day or night – or the movement of a boat will transform takers into the dourest of sulkers in a river pool or move the trout on to other ground in a lake. This latter is why so often the first drift will produce fish while the second and later ones do not. You may know you are on what is normally good ground, but in fact the fish have moved away.

Salmon in their sea-life are fish of deep water and it is understandable that in the relatively brilliantly-lit

PLATE
2

NORTH WEST WIND CLOUDS, HARD AND GLARING

PLATE
3

SALMON SEEKING THEIR NORTH IRISH RIVERS TRAVEL THE ATLANTIC TIDE-RIPS BELOW THESE CLIFFS

and shallow pools of a river they will be in the state we call 'shy'; they are noticeably less shy in a lake where the contrast with the sea environment is less sharp. Admittedly no more than a theory, this proposition cannot be made to apply to seatrout which are infinitely shyer than salmon in both river and loch.

Recaptures of marked fish show that seatrout differ in sea habit as their native rivers differ. We have fast-growing east coast fish from rivers of generous estuaries providing rich feed, and slower-growing races from the west coast where long estuaries are the exception, and downstream-moving fish go straight into the sea. In the one case most of the fish stay in shallow water feeding relatively near 'home' – in the other they go far afield in coastal waters. There is some evidence that many big seatrout travel with the salmon and like them are recaptured in deep water. They certainly are wanderers beyond coastal regions. All in all, however, it can be said that seatrout are mainly shallow water feeders and their sea movements in this respect do not take them into water so very different from that of the river.

Why then should they be our shyest game fish? I think, but I do not know beyond doubt, that their vision is better than that of salmon though this idea can be weakened if one considers that hour by hour and day by day their state of alertness in fresh water is greater. One seldom sees seatrout, however stale, in the state of torpor that sometimes possesses salmon – they always seem to have complete awareness, as do brown trout, of every movement and of every vibration.

I feel that in endeavouring to account for the seatrout's shyness, we have quite simply to conclude that they are surprisingly alert and keen-sighted fish behaving like the trout they are, and well-conditioned by their harassed and predator-haunted existence in that other life – the sea.

V

Days of the Black Dog

You can call him a black dog, a gremlin or even the gent with the cloven hoof himself; it will not make the slightest difference, because there will be days when he will sit on your shoulder and nothing you can do will dislodge him. But until you evolve some suitably powerful incantation to make him go, or he gets weary of plaguing you, there will be no fish in the boat. Instead there will be undetectable weaknesses in new casts and irons that break at the barbs, loops of line round row-locks when the only fish of the day is running hard, a dead calm over the best beat when it is yours and a lovely breeze when it is someone else's – one could go on and on. The boat scrapes on shallows she knows well and you realize she's low in the water with the weight of despair in her crew. Your gillie has been your man for twenty years but after a week of the black dog he begins to wear an odd look about him – at first resigned but cheerful – then puzzled and slightly furtive and finally you see doubt in his eyes.

You think he doubts *you* – wonders 'whether the old man's lost his touch' but it is not that at all. He is doubting whether all this can be chance and comes in the end to realization of what you have known since about the third spell-encumbered day. Once he is fully aware of the situation you have more chance. There are some country-bred incantations and procedures which he may know. One of my Irish gillies used to give the gaff a good hiding with a rope's end first thing in the morning – that has been known to work. Someone, I think it was H. T. Sheringham, recommended that at the end of one of these evil days, you should take solemn oath to renounce fishing for ever. This, apart from being rather soothing, could lead to first-class sport next day.

The best way of all out of it, of course, is to find a salmon that is really stupid – one that takes with a solid thump, sinks the hook deep behind a bit of tough gristle and then plays gently round the boat without snatching, jigging or in any way imperilling the tackle in which by now you have no faith. Get him in the boat, tap him on the head twice with a Guinness bottle (and once again just to be sure) and your troubles are over. The thing that has been sitting on your shoulder does not like to be thwarted and he will depart as soon as the fish is laid in the boat. It need not be a big fish – a grilse will do. I have seen a trout work effectively too but in this case it *has* to be a really big one – the kind of which admirers will say, 'I'd rather have this than *three* salmon'.

However wide your experience, profound your knowledge or deep your cunning, you are certain at intervals to attract the attention of the black one. Just keep on fishing.

I had a visitation one recent July while seatrout fishing in Ireland. The first manifestation of something amiss was devastating – the seatrout had not arrived though in early July in the average year that lough receives a big run. The fish were there in the estuary four miles away, going up and down with the tides but not, in general, attempting to run the very low water. A trickle came through and we had a few fish, a very few, each day, fish in the one-and-a-half to two-pound class. But they were difficult to find and we covered a vast area of normally very good water between rises. Those we found rose and took well and as always, a number nine or ten Golden Olive was the magic. At that stage I was no more than disappointed; this kind of thing happens at intervals through the seasons and is all part of the game.

The true state of affairs dawned on me, however, when I rose a salmon and, seatrout attuned, smartly took the fly away from him and when, next day, my partner in the boat, relatively a beginner, boated a salmon. It had obviously entered on the last little spate two weeks previously and was pretty fresh. It certainly would not have come alone and there must be others of the same class. So, while not abandoning seatrout altogether, we began to fish areas which are also salmon ground. Here the black dog on my shoulder really began to work. In the next four days my companion had a second salmon while I did not have a single rise, my only contribution to the bag being a two-pound seatrout. We were fishing some of the finest salmon water I know – great holding ground between entering and outgoing rivers – four or five feet deep over sand, gravel and boulders – and the fish were there. True, the breeze tended to be light but we had spells of a big soft swinging wave and good light and I was looking for a fish at every cast. It was like that on the fifth day when we pulled away from the beach after lunch to get an offing for the first drift. Sometimes fish, both salmon and brown trout, will take a fly thrown at right angles to the boat's course and allowed to swing back into the wake, coming to it usually just before it begins to drag. I was casting thus and pretending to be careless to deceive the black dog, when a salmon took the fly with a great splash and a heavy pull. The way he took hold and his subsequent behaviour showed he was well-hooked and while he went off in a longish run across the wind, the boat was worked off into deeper water. Presently we had the fish circling near the boat, just under the surface, a nice clean-looking fish of twelve or thirteen pounds with the dropper, a six Silver Grey, in the corner of his mouth. A little extra strain and he came, tired and rolling, towards the outstretched net. A foot away he gave the very slightest jerk of his head, my black dog reached out a paw and the fish was gone with the fly. The inquest showed that the cast, new that morning, was necked at the eye of the fly. I admit I had not examined the cast for an hour or so but there had been nothing to suggest the possibility of necking. A gusty, awkward wind will do it as it will, occasionally, produce wind knots but there is really not much chance of either if cast weight and fly size are balanced as in this case and the breeze is steady.

We went on our way, the gillie helpfully recalling some words I had forgotten. But I knew now, beyond doubt, what was the trouble and I was not at all surprised that it took us two more days to find another willing fish. This one rose beautifully and went down with the fly giving a deliberate purposeful wag of his tail as he did so. On reflection I think it may have been a wag of derision because after sailing about rather sedately for perhaps three minutes this fish very gently came unstuck and the fly came back with that sickening weightlessness. I could not think of any words to suit the circumstances so we treated this episode as a joke. After all you can lose a fish this way at any time – nothing unusual about it. But I knew. . . . There had been the unaccountable disappearance of a box of flies just when I needed the patterns in that box, my missing a ring when threading the line on the rod and having to take the cast off and do the job all over again that very morning, and sundry other small bedevilments to boot.

But we kept on fishing.

Three blank days later – on the morning of the fourth to be precise, we had the best conditions we had seen. Fishing a southerly shore of rocky points and boulder-filled bays in a strong steady north wind we had a big wave so that it was just about all the gillie could do to hold the boat. These are ideal conditions for this type of water and despite the Jonah in the boat I had great hopes. I must tell you that I had resolutely refused to change boats; it was a great temptation because a boat change sometimes works wonders. On the other hand, a certain strength of character must surely earn *some* reward?

Well, you shall hear of my reward. It came in the form of one of the oddest experiences in a long fishing life and I do not expect it ever to be repeated. Since salmon move down-wind in a big wave and blow, they are very often found, in such conditions, in the breaking wave zone – say in about eighteen inches of water among the stones of the beach. They move to and fro in the oxygenated water and will take well if you can find them by fishing a long line from a boat held well away from the taking area.

Doing just this I moved a fish among some rough rocks in a couple of feet of water. He took quietly without breaking the surface, gave two or three rather ponderous shakes and the feeling of being well-hooked. He also felt like a very heavy fish and I held him with firm pressure until he should move off or do something. James was beginning to get the boat's head into wind and wave, in anticipation of vigorous activity of some sort, when the odd thing happened. The fish came to the surface, not as usual on an even keel, but flat on his side – we had a full twenty seconds to take a good look at the biggest fish we had seen in the lough – certainly twenty-five pounds and perhaps nearer thirty – bright and pinky-mauve and very beautiful. The line was tight but not too tight and the fly was not visible. Fins and tail moved almost lazily in the wave and quite gently he arched the whole length and breadth of him and slid away to leave the fly streaming in the wind. I have never seen a fish lie flat in this way – though a freshly-hooked salmon will sometimes stay motionless on an even keel for a time – and what caused this big fish to do it I cannot explain. His loss was really the very end; if thoughts are shaded mine were black as the dog so firmly on my shoulder, though I fished hard, without result of course, for the remainder of the day.

That night the wind shifted south and because of the change, in the morning we had to be on different ground and in a very light breeze and a ripple that was ideal for seatrout and too small for salmon. I fished, I confess it, with little hope, not carelessly but rather resignedly and presently had a rise. I saw the surface break and a quick flash of brown that might have been a two-pound trout but was, in fact, a fifteen-pound salmon.

We looked at it, James and I, as after a time it lay in the boat. Fish scales on the floorboards at last, and the lough lay fresh and smiling again under sunlit mountains. Black dogs indeed – none in this boat!

Just keep on fishing.

VI

Why Does a Salmon Take?

WHY does a salmon take a fly or a bait? That question was put to me quite recently and I was rather taken aback – first by the difficulty of giving any reasonably brief but still adequate answer; second by the fact that I do not know any more than anyone does; and third by my own astonishment that after volumes have been written on this one point, anglers should still ask the question. It is always in every salmon fisher's mind, as it has been since it was appreciated that salmon do not feed in the river, but one so seldom nowadays hears it put into words and not much seems lately to have been written even in speculation: no one takes the trouble any more to summarize our (lack of) knowledge. Attempts have been made of course, to relate the form and colour of the fly to those of the sea creatures on which we suppose the salmon to feed and with which the fish must therefore be familiar. Arthur Ransome, that charming writer and good fisher, did it with his Blue Elver and his studies of flies to simulate the larger elements in the plankton of the sea, and others have developed theories and lures to suit them. All these fishermen put their strongest faith in one conception of why a salmon takes a fly. They may easily be right. None of their special confections, however, to my knowledge attracts salmon consistently better than some other and older patterns which do not fit the inventors' theories at all.

I can contribute no new basic knowledge on this subject, nor I suspect can anyone else, but the idea I have just touched on is only one of many and it may be that there is something to be gained from discussion of this, surely the most enduring of a salmon fisher's curiosities as one generation follows another on river and lake and puzzles over its failure to interest the fish – or over the reason why fish *do* take. I am encouraged to proceed because when I suggested in rather deprecating fashion some of the old familiar reasons why, supposedly, a salmon takes a lure, the man who had so surprised me by putting the question became as interested as I would have expected had the answers been entirely new, and begged me to argue it out. So I propose to do just that, leaving blasé fishermen to pass over this chapter, conclude that they had not thought of this or that or condemn me – according to their sophistication, charity or belief.

Since we began by touching on the belief that salmon are interested in certain forms, colours or silhouettes of lure because they simulate something which the fish has known in its sea environment we may as well continue along this line. The belief itself appears reasonable if we assume that a salmon will take something, whatever it may be, as representing food. The familiar, in the food category, would surely be accepted in preference to something unfamiliar. Most creatures, wild or domesticated, will accept the one whilst refusing or being very hesitant about the other. What then, of the undoubted fact that salmon do not feed – which means for our purpose do not take in solids – whilst in fresh water. Well, we have to distinguish here between ability and inclination because the fact of not feeding does not by any means destroy the contention that a 'memory' prompted *inclination* to feed may be present in the fish. In short, the sight of a moving object resembling some component in the fish's sea diet, may be all the stimulation required to provoke what we fishermen call 'a take'. I think on balance I am inclined to favour this idea over the several others which are at least feasible.

Although sea-caught salmon are not gutted as are most other fish so that the stomach contents are seen, we do know the broad range of their diet and that clupeoid fishes, the herring and its relatives, form the bulk. Deep-water prawns and the crustacea, perhaps small octopi, play a part and are largely responsible for flesh colour – I am not persuaded as to the extent of plankton feeding. I think the salmon appears here only as a predator of the herring kind that feed on the plankton. Closer inshore, sand-eels certainly are freely taken and the common prawn and shrimp of these waters too.

Does the salmon bring its habit of pursuing these forms of life into the river regardless of its inability – arising at some stage during its riverward migration from far-away feeding grounds – to assimilate them? It seems likely – a small spring fish has been about two years in the sea and the habit must be strong. There is something else which has always seemed to me of some significance – and that is that as river stay lengthens through the summer, the fish, now in its other environment, and with sea habits months away, is more and more inclined towards interest in small sober-coloured things – which it pursued from habit in parr and smolt years. Is it that summer life in a river pool full of trout taking small nymphs and fully-fledged insects – in the shadowy weedy world of the salmon's first years, can re-awaken a habit? Most fishermen have seen salmon taking natural flies along with the trout.

One might argue against this that in the autumn when fish have had longest to regain their river habit, they are more inclined to take larger and gaudier lures – not smaller and quieter-toned ones. This would argue a re-growth of sea-habit (if the sea-habit idea is sound) and that I think is absurd. It is much more likely that salmon in autumn take whatever they will come to, for quite different reasons. Before turning to those reasons, however, one ought to round out any sea-habit appraisal by looking at some of the lures that a salmon will take in the river. Spun baits, of course, are legion and in suitable conditions – mainly heavy and/or coloured water – are freely taken. In general artificials can be said to possess small fish characteristics in shape, flash or colour and action and, so far as we know, are taken by the salmon *as* small fishes. Natural baits obviously are. Natural sand-eels can be very effective and of the many salmon I have had on this bait I can recall one particular fish on the Taw which, having refused sprats large and small, gold and silver devons, plugs and assorted spoons, took a plastic-coated natural sand-eel at the first cast so savagely as almost to take the rod from my hand. This fish had been not more than twelve hours in the river. This incident by itself means nothing but the sand-eel – identifiable sea food of the fish – undoubtedly has some special attraction.

Hundreds of salmon every season take the worm or a bunch of them and I have heard and seen it suggested that the fish may feed in the sea on small octopi and take the wriggling clot of worms because it looks similar. Informed comment on this proposition is not possible as far as I am concerned but my feeling is that the octopus, large or small, is hardly likely to be any sizeable item in the diet of salmon which do not feed beyond the limits of the temperate and cold seas. There may be other forms of marine life on the salmon's feeding grounds which in form, movement and coloration, are reproduced to the eye of the fish by that trundling bundle of writhings which is taken, crushed and eventually ejected in fresh water.

Two forms of marine life without guise are offered to and taken by salmon in fresh water – the prawn and the shrimp. There are puzzling features about the reaction of the fish to the prawn but it does seem reasonable to suppose that both baits are taken because they are recognized by and familiar to the fish. Some forms of deep-water prawns are as highly coloured in nature as they are when dyed by the angler but I suggest that the bulk of those consumed in salt water are not. Sometimes a fish will take a deep crimson prawn – another time a golden one. The preference may be related to light and water clarity; it may on the other hand indicate that form, shape or silhouette play a greater part in recognition than does colour. The sense we call taste is another factor – but it hardly fits our context since it only comes into play after the fish has 'recognized' and taken the lure.

What is very odd about the prawn is its excessively *unattractive*, indeed repellent property, at some times.

If, as this discussion requires logic, the prawn is taken because it has been the sea-food of the salmon, does it follow that at some time in the sea the fish find it as repellent as in the river? Anyone who has seen salmon retreating from the presence of a prawn fished through a pool, can have no doubt of their dislike for it; when they take it they can do so quite gently or so savagely as to yank the rod tip down! It is when one comes to reconcile one facet of salmon behaviour with another that the possible range of permutations begins to take shape. As one introduces the unending variety of fishing conditions as backgrounds to this or that fish reaction there emerges all over again the vast extent of the anglers' ignorance of the incomprehensible salmon.

Flies designedly simulating forms of life from the food range of the fish are not very numerous, apart from the several dressings based on the traditional 'Prawn' fly. It can be said that translucent and hair-winged patterns in the style of yellow and orange versions of say, the Torrish, form very passable imitations of prawns and shrimps as the salmon may see them especially when the fly is bleached by repeated use. The only one, I think, in outright and really good imitation of a salt-water item is the Blue Elver. This fly, wet and working, long fibres undulating with current pressures and pull from the line, is extraordinarily lifelike. I criticize it as a shade bulky but it certainly moves fish and particularly fresh fish. I would think that seventy-five per cent of the fish I have had on the Elver have been newly from the sea though it is not, in my experience, notably effective in sea-pools affected by the tide. When the sea-pool, so-called, is really only the river-pool nearest to the sea and not brackish, the Elver can be as deadly as anything you can fish.

There is a distinct parallel between the prawn as a bait and the elver in its free natural state in that sometimes the natural elver like the dyed prawn is shunned by the salmon as if it were the plague. I have never seen a salmon take a natural elver but since the fish take the artificial I am prepared, even inclined, to accept that they do feed on the natural elver in its immature (*leptocephalus*) stages in the ocean or its 'mature' state as an elver in coastal waters. But I *have*, and very often, seen evidence that salmon abhor the elver when it is concentrated in its myriads in the river. It is said that the little wriggling pests find their way into the open gill-covers of the fish and that the fish are hence terrified when their resting places are invaded. I cannot personally vouch for this but I can testify that I have always failed to find salmon in pools filled with elvers during a run of the baby eels. After counting the salmon lying in small bare rocky pools in a low June river I have returned next day to find a big run of elvers in progress and the salmon gone. No other factors were at work, the river was too low for fish to move up of their own free will and there was very little cover for the fish. I can therefore only think that they had found refuge in such white water as there was and where the elvers could not make headway. As a habit, the elvers tend to move up the margins and through quiet water and will wriggle up stones which are out of water and in fact anywhere so long as it is damp or splash wet. The example I have quoted is one of many and I am quite certain that masses of elvers and resting salmon are incompatible. Perhaps, then, any willingness to take the

PLATE
4

THE GILLIE HERE SCORNS THE BOAT'S MODERN NET
AND USES HIS OWN FAVOURITE

PLATE
5

PLAYING A FISH

elver as a food-resembling lure is dependent on the number of elvers present?

The observant fisherman who will realize that the things he sees while he is fishing are a part of fishing and is therefore prepared to turn aside from casting to look and think and marvel, is a happy man. In this connection I cannot leave the subject of elvers without some comment on the miracle of their running. I first saw them swarming upstream in a highland burn when I was a very small boy and I have seen them in rivers of the north and west in most years since then – never failing to be strangely moved. Putting aside the wonder of their ocean journey – there is something which I put with the tremendous events of nature in the arrival of these very small creatures from the sea. You sit by a small steep-coursed burn in a remote moorland and there they are – black ropes of them on every wet surface, going vertically up the mossy rock. At the peak of the run a dipped bucket would bring up as many dark undulating needles of energy as will make more bulk than the water. Their progress upstream is surely the ultimate in collective purpose; they are so fragile and the odds against them so great. As a fisherman I have no use for the feeding and growing eel but I cannot watch the elvers going upriver without utter fascination.

Apart from this digression we have been dealing fairly consistently with one of the reasons why it is thought salmon take a line – the food familiarity theory. What of others? What, for example, of the argument that they take out of curiosity or irritation?

Both 'reasons' have the merit that they can be made to explain the salmon's occasional willingness to take objects so grotesque, to our eyes, that they surely cannot be held to resemble anything the fish recognizes or treats as natural. Many flies and some baits so qualify. The curiosity angle is, I feel, a weak one. It is almost a part of the food theory and pays at least lip service to it. Is there such a thing as pure detached curiosity in a fish? I think that the impulse is infinitely more likely to be prompted by some food similarity showing to the fish in a way which we do not appreciate. It may be in movement or silhouette in certain lights rather than in coloration or true form. And it must be admitted that since our knowledge of the salmon's sea diet is limited, there may be items in that diet which match our more grotesque lures.

The irritation motive cannot be dismissed lightly. We know that salmon will and do display a reaction, which in all the attendant circumstances we interpret as irritation. In our eyes the fish behaves as though irritated – coming as fast as a trout – unnecessarily fast – to a slow fly and taking with a sharp snatch. Or taking after it has been shown a succession of lures patently unattractive for some reason we may speculate about but do not know. When such a fish takes, we conclude we have *teased* him into taking; he has finally snatched the fly out of irritation with the impudent small thing which keeps passing before, over, or beside his lie. Sometimes, of course, he refuses to be irritated – he just lies unmoved and the more he is teased the more recusant he becomes. Like most salmon fishers I flatter myself that often, in fishing for a known fish in a lie I could pinpoint, I have evoked an 'irritated' take, but in my heart I am not sure that I put much weight on this idea.

I would rather bear in mind the behaviour and reaction to an intruder – however small – of a salmon, especially a cock fish on, at or near the redd. The urge to deal with any intruder is clearly one that develops as the fish is re-absorbed into his river environment. His lie for the moment is a small world of which as his gonads develop he is increasingly aware. In the hours in which he is alert he will take an interest in and examine to the extent of taking in his mouth, anything which moves other than drifting or crawling minutiae. If you want to categorize this sort of behaviour under the heading of manifested curiosity or irritation, you can do so. It can be one or both operating as part of some defence mechanism which is itself a reason why salmon take some of the creations we offer them.

A brief summary on this subject is impossible and I do not propose to attempt one. I think it best to add simply that the evidence of my own experiences leads me to have most faith, if I have to make a choice, in the theory that salmon take because they see something that awakens a sea-born feeding impulse.

VII

'Difficult' Fish

IN THIS age of specialization, even in angling, there are still surprising numbers of 'chuck and chance it' people. One of these whom I had invited into discussion of fishing problems he encountered was forthcoming in a most disconcerting way. I had picked the wrong angler because unexpectedly but probably deservedly I had the answer, 'Tell me how to catch fish!' which was fair enough but not the reply I sought. I did not wish and will not try to suggest solutions to basic problems which arise from sheer lack of study and application. An assumption of considerable proficiency in the reader is inherent in these writings which seek to deal with the refinement of problems and not really with the basic elements of fishing.

It appears to me that the kind of situation I had in mind can fall into one or other of two broad classes. They can concern the difficulties of a set of conditions which affect the behaviour of all fish subjected to them on a certain day or over a period (and many of these pages are devoted to them) or they can cover the problems created by the situation or attitude or character of one specific and defiant fish. If it is not easier to deal with the latter category in the sense of offering explanation, it is at least easier in the telling by way of illustration of experiences in trying to deal with awkward individual fish of the salmon kind.

Take, for example, a salmon whom my ghillie and I and others, came to know, with feelings of affection, irritation and amusement, as 'Old Joe'. He lay close to and sometimes among some jagged black rocks in a very small shallow bay on the southern shore of a wild gale-battered loch and here one rough September day he came charging out at the fly, seized it and was gone with a prodigious splash, apparently well-hooked. He came out of the water ten yards away and hung suspended in mid-air long enough for us to get a good look at him, to recognize him for a very red cock fish of about fourteen pounds distinguished by an unusually big tail and a jib that would have done credit to a November spawner. We also saw the fly drop away from him.

These ripening cock fish are often aggressive takers and we expected to move him again when he had been given time to get back to his lie. So we pulled out of the bay and away to another drift and back an hour later. I went up two sizes in the fly but kept to the pattern – a Black Doctor – on the theory that the bigger fly had a better chance to grip somewhere between those gaping jaws. He was still there and the fly had not moved a yard before he had it with a savage pull and if that was not enough, I laid into him at once to sink the hook point into that hard jaw. He felt rock-firm and surged off on the surface with back awash. Fifteen yards of this with a firm strain on him – and the fly came back! We could not, of course, see what happened and I have no idea just where the fly had lodged at the moment of taking, but clearly the big fly had no more hold than the smaller one.

Perhaps a double might do better so a double Thunder and Lightning (the only double of suitable dressing I happened to have with me) went on and we tried again almost at once. The fish had moved a yard or two from the rocks but he came again as eagerly as ever and treated us to two of his splendid jumps. I thought that when the hook held through the first jump we had him but it came away in the second and the fly was clearly visible coming out of the inch gap that was the jib's effect in preventing him from closing his mouth. But why, when he took so very well, did not the hook, especially the double, go further back in the mouth and take hold in the less leathery area? This was and still is, something of a mystery but we were interested and amused, having had an afternoon's fun from a single fish.

When I was questioned as to my plans for the following day I said that I was going to catch 'Old Joe' and I did try. I tried very hard indeed and incredible as it doubtless seems, I did hook him again and precisely the same thing happened – the fly simply dropped away from those hard brown jaws. I think even 'Old Joe' had had enough then of the strange wavering creatures that kept invading his lie for he refused to come again. All I could think of was changing the angle of presentation and this could not be done because of the rocks that littered the bay. Two friends of mine came back from the loch a week later having had a day's fun with Old Joe, but he was never caught. He flourished many years ago and today with a fly dressed on a small fine-wired triangle in the modern fashion, I think I could hook him properly.

Whatever form the fly took, that fish would at least take it, and one had hold of him if temporarily, and to that extent he was less tantalizing than the fish which utterly declines to rise or worse still the fish that rises without touching the fly. This latter can appear anywhere, at any time and most fishing days provide individual examples. But I know a place where all the fish do just this and oddly enough it is not half a mile from the scene of my frustrations with Old Joe – in fact in the same loch. The immediate stage is rather different because here the lie of the fish is the shallow, bouldery water off a long, low point of heather-clad rock. The fishing area which makes up the so-called taking place is ten yards wide by twenty long – a narrow rectangle with one end on the tip of the point and the other at sea. The water shelves very gradually from the shore end to two feet six inches or three feet and really covers one of the shallowest lies that I know anywhere though I have caught many salmon in water just as shallow in other Scottish and Irish lakes. These, however, have been isolated fish, not a concentration of them. Over the years I suppose I have risen thirty or forty salmon on this point – and killed one. I do not really know why results are so poor; I feel only that the extreme shallowness has something to do with it though why this factor should operate here and not in other places remains a very pointed question. What happens is this. Because of the thin water, one fishes a line slightly longer than usual and the boat is kept clear of the lie as long as possible. To fish it out, the boat has finally to intrude and after that it is finished. There will be no further rises. Every fish risen, and on a good day with a fresh stock in the loch you will move four or five, comes with a lightning-fast splashy rise. The water explodes under the fly – whether tail or dropper, one waits for the line to draw away – and there is nothing! You cast over the fish again at once or pull away – come down on him again in five minutes – perhaps come down in fly size, all according to your fancy or personal practice and he *may* come again in the same manner. More often than not he will not come twice and you seek another. The same thing happens – a fussy, splashy rise and no contact. 'Aha,' you say, 'no use treating these fish like salmon – these quick risers should be struck like seatrout.' So the next one is treated as if he were a seatrout in that fish's quick-taking mood. That does not work either. You try fishing very slowly indeed, giving the fish time to take quietly – they do not like that – the total absence of rises prove it. You ring the changes on fly sizes until absurdity at both ends of the scale is reached and that fails too. These fish will come to the right fly in eights or sixes according to wind and wave. They are attracted all right but they do not take hold.

I can hear the reader muttering and I do not wonder. But a number of others who fish this loch oftener than I do and should have the feel of its moods better than I, have also failed with these fish off the point. Elsewhere in the loch, fish take the fly well and it is by no means what is called a 'bad taking' place. The trouble, which I associate with the extreme shallowness, is very local. The one fish I have taken here seized the fly in less than a foot of water and without a fraction of pause, ran very fast, with the reel screeching, for about forty yards before doubling back on a very slack line. But we boated him. I thought at the time that we had at last dislodged the Jonah that sits in every boat fishing the point but it was not so. Difficult fish indeed.

There was, once upon a time, another single fish that beat me, as 'Old Joe' did, but for much more obscure reasons. I rate the incident as the major failure in a long pursuit of fishing and I call the victor 'the fish at the nine-eyed bridge'. A nine-eyed bridge, in the western Ireland where it is located is, as might be supposed, an old stone bridge with nine arches. Some of the arches are choked with earth and scrub but the rest carry a narrow valley road across the river, here a fast stream of dark peaty water joining two loughs. The lower lough and the river yield large numbers of fish to the fly. One late April day, with the river falling after a spate, I fished a pool just above the bridge. It is an odd pool of which the lower third is unfishable because of a big ash tree which, uncut in this neglected place, hangs down so that the lower branches droop like a curtain across the stream, tips in the water. This is a pity because the tail always holds fish and the body and neck far fewer. The bank is fairly high and rough but the open part of the pool fishes easily enough. I came to it with high hopes because fresh fish had come straight through lower river and lough from the sea and I had one bright ten-pounder on the grass in the first half-hour.

The top run was fished and the centre section of the pool without result but as, ten yards above the ash, I allowed the fly – a number six Thunder and Lightning – to work across just upstream from the trailing leaves, a fresh-looking fish of sixteen or seventeen pounds tilted rather deliberately up and took the fly, hardly breaking the surface. I saw the fly engulfed and the cast leading from the corner of the mouth as the fish headed upstream without the slightest sign of turning down or to either side back to his lie. He passed me just under the surface and I tightened when he was well above me. To my absolute astonishment the fly came back. If I had not been there with a slack line dangling from the rod tip I would not have believed it could happen. Though the incidents with 'Old Joe' had not then taken place, I am able to say now that there was no question here of inability to grip the fly – this fish was not twenty-four hours out of salt water and I had seen him close firmly on the fly.

Well, reflecting that it *is* possible for a hook to be flat in the mouth of a fish and so fail to hold, I tried again after a suitable wait. The fish had dropped back to his lie and came up again in precisely the same place – two feet upstream of the ash tree tips and precisely the same thing occurred. It still is incomprehensible to me that one could exert a downstream pull on a fly somewhere inside the mouth of a salmon without pulling it into the scissors where it must surely hold. Beyond understanding it may be, but it is certainly possible because it happened four times in all before that fish lost interest. I did not change fly size because I could see no contribution from that but I did try fishing beyond and below him to bring the fly up past him and perhaps make him turn to take it: that did not work. He came to it only when it swung slowly across and above him and he took it straightly without turning.

Again I think a sharp-hooked triangle, not then devised as arming for a fly, might have held him.

The three 'difficult' cases I have cited are so very unusual that my fourth should refer to a situation which of its nature defeats most anglers, myself included, and is often met – a situation in which salmon seldom or never give one the opportunity to hook them because they simply do not rise. Now there are days when fish refuse to rise in holds normally classed as good taking places: we all know such days and the conditions that give rise to them, but there are also places where fish seldom take, though numbers may be present, even in good conditions. Unending volumes can be and have been written about the first kind of circumstance because it always embodies an element of hope but the second is very different. Its significance in a fishing life rather depends on where you fish because it is most evident on one type of water and though I know bad taking places all over the British Isles as well as farther afield, I find the trouble worst in the boglands of the west and north. It is so often a problem of the peat river which drains high ground to meander in the lower reaches through peat-bogs and flat wet moors. Here the river cuts deep into the peat through narrow twisting channels. Banks may not be high but the ribbon of water gets no wind unless directly up or downstream on a straight. Generally you fish in patches of light ripple or calm. I could list any number of such places, canal-like currentless stretches at times full of fish; they are like other pools in one respect only – there are usually one or two places where in extremely good conditions, high water and a full gale, you may move a fish. They have little to distinguish them – it is always 'just by that tuft of rushes' or 'just where the ditch comes in'. I have had a few fish from the rushy tufts and the ditch ends in my time but it is so difficult to get good water height, and wind and wave just right that they have been few indeed. Yet I have seen these 'pools' swarming with salmon. One I know well is a hundred and fifty yards long, deep and black and almost wholly without current but with a dog-leg towards the tail. Below the bend is a widening with reedy edges and a lily-bed and here the fish congregate to show at intervals on most days – splashing, lunging and swirling in the calm water. Most of them are black: they start to colour up at once on entry with the peat stain. The inexperienced linger here casting fruitlessly among these exasperating salmon – the old hands know the place and leave it alone until that gale comes along and great waves run down the pool. The wind blasts the moor and if it is upstream you cannot get a line out across and down. But you *can* fish upstream, and this way I have had a fish or two. There is a very faint hidden current at most water levels and a fair one in a spate. An upstream wind then raises short steep waves and in this the odd fish will come to a small dark fly – a Butcher or a Black Doctor – as small as a nine or ten. In average conditions these fish will not rise at all and I do not know how to catch them. It is a pity because there is a great deal of water like this.

VIII

The Fourth of the Month

THE little river rolled brown into the western sea across its shingle bar, and in the long flat above an odd small salmon made a plunge. There was no sign of seatrout but I knew the river would be alive with them. Twelve years of life – six of it fishing – had given me the stirrings of knowledge about small rivers. Among other things I knew was that a heavy bag could be made with the worm fished in the quieter margins and little current-sheltered pockets of that brown river and I itched to be doing it. The deterrent was not any absolute prohibition but partly parental disdain of a 'dirty business' and mainly the comment 'Any fool can catch them in thick water'.

It would be idle to pretend that by inclination I was particular, indeed I was not above guddling (tickling if you like) a few trout in low water but I liked to be, and to be seen to be, a fly-fisher. And so I sat on the rabbit-cropped turf by the river-mouth and watched the river in flood from rain in the hills, now over. Tomorrow it would have fined down and I would be out early to fish the fly. I was. I was out before breakfast, kicking up hordes of scurrying rabbits from the bracken, working a bare hundred yards of river above the beach and returning, morning air hungered, with four small seatrout all, I can remember, on Teal and Yellow. The kitchen incurred my displeasure by their totally unwarranted desire to dispose of breakfast before preparing sandwiches but eventually I was back on the river with a whole summer's day before me.

The water was no more than misty by now and there seemed to be seatrout behind every stone and in every run and under every undercut turf bank. They were all fresh and glittering and soft-mouthed and I pulled the fly out of some of them. There was a clump of old whins by a little footbridge and here in the cover of a bush split apart and lying on the turf, I dumped about a dozen fish. My recollection is that they were ten ounces to one pound with a giant of one and a half pounds to dwarf the lot.

I can see that piece of river now and the place much higher up where in the late afternoon a local boy of my acquaintance came running down the farther bank. He waved a piece of paper of some sort at me and shouted 'War declared on Germany!' It was August 4th 1914 and I just went on fishing and making little caches of fish under heaps of bracken here and there. My frayed cast parted in a fish and I had to wait while a fresh one soaked with a stone on it in the river margin. The stream was becoming small now and increasingly hard to fish with the fly but I still caught fish, and lost count of them.

The details of all that day, even if I could recall them now, would perhaps be wearisome. It will do to say that I took the rod down when something told me it was long past tea-time. There could be no short cut home because I had to pick up the fish I had deposited here and there on my way upstream and when eventually I arrived, I had fishing bag and every pocket full and I was weary with the long day and the weight that hung inconveniently around my person. A second visit had to be made to the two caches nearest my starting point and then the counting began before an astonished household: with the four from the early morning I had forty-two seatrout averaging about three-quarters of a pound. Numerically it was then my best bag and numerically, all these years later, it still is.

Even by weight it rates among the better bags for one day's fishing. The many brownies were returned with the exception of one fat yellow giant of about a pound. It was a great day and I shall always remember that sweet little river with affection and without it the cry of war across the water.

An infinity, it seems, of years later and indeed not so many years ago, I had another quite unexpected good day that was also the fourth of the month – June this time.

I had gone to the Hebrides to catch brown trout which take well in the islands in an average June. The fish are plentiful – too plentiful – and consequently usually small but there are some lochs with a surprisingly good average weight, when one considers the type of water, overall scarcity of insect and molluscan life and the very short season when airborne food is (relatively) plentiful. The innumerable little dhu lochs on the moors are full of little black trout but there is an intermediate class, rather larger lochs as a rule, which will yield what on the mainland we used to call 'herring-size' fish and then there are a few in which trout may average just a little better than three-quarters of a pound with the odd big fish – very hard to find – running to five pounds or more.

There are, of course, a few places in the long island

where there is an early run of salmon and a few seatrout but July and August are the months and a June visit is too early for migratory fish. One goes, as I had gone on this occasion, with the determination to be content fishing for brownies. I can be very happy with a little rod and light tackle in a boat on a loch full of trout and I recall with gratitude some glorious days of Hebridean June weather doing just that.

First, however, it has to be explained that the principal scene of operations was a big irregularly shaped and rock-filled loch at the head of a long chain of others in typical island disarray. Short connecting rivers in late summer brought salmon and seatrout in from the tide. It needed big water to bring them to the top of the chain though by October the bulk of the salmon would be there. The loch is bejewelled with islets and abounds in gravelly shallows and sandy bays, but in all the miles of shoreline there are only three small patches of weed. Trout vary widely in coloration and condition – those from the dark soft peaty shores match their background, those from the commoner clean stony areas are well-conditioned, shapely and golden of belly. They are free if spasmodic risers and persistent fishing could produce very big bags. My best has been sixty-one trout for a day beginning at noon and ending in early evening and I think a full day in the right conditions could yield a hundred. Of course, it is the practice to kill only the best of the fish.

The day I remember so very well began with a three-mile tramp up and over a high ridge and down to a long straggling shallow loch, to be traversed by boat. Then, that boat drawn up again, over an area of peat gullies and old deep heather to final embarkation point. It was bright and warm with a brisk southerly breeze and a few soft clouds to give us a changing light. I had Angus for gillie and he is one of the two best I have ever fished with, a very light nine-foot rod, a small reel with thirty yards of casting line, a little backing and a 2x cast with Invicta on the tail, Watson's Fancy on the dropper and Butcher on the bob. To judge from the water-level ridge on the sands by the boat, the loch had lately been a foot higher.

Trout began to move as soon as we were out of the calm lee of the higher ground and we drifted at precisely the right speed across the bay. The water sparkled, a pair of red-throated divers watched us intently from a safe distance and cloud shadows chased across the green of new spring grass on nearby hills. The colours dimmed as the hills receded until the tips of the Skye hills were the faintest of blue battlements on the far horizon. We had the beginnings of a basket of trout by the time we drifted into deeper water and they ceased to rise. Close to shore again every little shallow point produced a fish or two; we kept only those approaching the pound. There was a crisp little wave that glinted in the sun and the Butcher on the bob exactly suited conditions. Angus has little patience with trout fishing but even he sat happily in the sun while I fished the flies very fast across small waves and got enough eager rises in return to keep us interested in fishing.

Where the bay opened into the main body of the loch was a long spit of stones and we pulled out from there to cover this very good shallow strip in about four feet of water. One good trout came from the bay side of the spit and then two small nuisances as we crossed what should have been the best water. And here, to my surprise and Angus's delight, a salmon appeared in a quick flashing interception of the tripping Butcher. If I had not seen the bright gleam of the fish as he went down I should have taken him for one of the big trout – on an even keel he looked brown and boldly spotted. 'If he makes a long run, we're done,' I said – 'so we had better be very gentle with him. If we get him we'll make history.' Salmon are not taken in that loch in June. Gentle we were, Angus creeping out from shore to safe depth while I held the fish firmly but not hard, on a fixed length of line. He went steadily out into the bay with no attempt to run or jump and there began to circle. It took twenty minutes to get him up but the unvarying firm hold did it and he came smoothly to the net without a head shake or a lash – perfectly fresh and about eight pounds. The first salmon ever to come from that loch in June.

There might be others on the spit off the headland, so while I looked out a heavier cast we went back. I changed the cast to one of 1x and increased the fly size to nine, eliminating the middle fly at the same time. The trout were not put off by the heavier cast and this time over we had two keepable fish but our salmon appeared to be the only one in that particular lie. Oddly enough, even when later in the season there are many salmon, that place is not a recognized lie. We concluded, however, that if as it seemed, high water had brought in some salmon it would pay us to fish some of the known lies and forget about our trouting. I longed for a rod with rather more power and a reel with ample line and backing. There is nothing like fishing with confidence and I dreaded the smash that would certainly result if we hooked one of the wild fish that make prodigious runs, taking too much line before the boat picks up speed. It is true that one kills them sooner but thirty yards of line is grossly inadequate for fish that commonly make sixty- to eighty-yard dashes along an unobstructed loch floor and finish that with a swift upsurge and a somersault in flying spray.

We tried a good beat behind a high heather-covered island and drew a blank. A narrow rill of alkaline water came down the mountain and into the loch here, and the stony bottom was dark with moss growth, the trout were dark but shapely and well-grown. Over lunch on the pebbles of the island beach we found we had eighteen, which was fair enough since latterly I had been trying to avoid hooking them. Somewhere among the rank heather and wind-stunted rowans above and behind us was a nest of vociferous young hooded crows and Angus, with a thought for his grouse and mallard eggs, made an effort to reach and destroy the nest and its contents but soon gave up. The island was a jumble of overgrown loose rocks with leg-smashing cavities and other hazards. I always know that when Angus starts to make little post-lunch excursions he is itching to be fishing again and thinks we have wasted enough time. So we collected the gear and pushed off again.

About quarter of a mile away upwind the loch shores close in to a narrow waist by reason of long lines of rocks which reach out from both sides, one to terminate in a small green island, the other in a pile of great boulders. Between them is no more than forty yards of open water – the only passage for a boat. The rock-strewn shallows on either side probably hold salmon but are unfishable in a light breeze and downright damaging to a boat in any real salmon-fishing wave.

Today we left the shallows alone, to concentrate on a few known lies on either flank of the channel entrance. We pulled into it and drifted out. The left hand side with a huge and notorious sunken boulder on the corner was fished without result but on the next drift and on the opposite corner a fish came up very quietly beside the fly, showed his shoulders and back as he did so and went down with an unhurried little wag of his tail. The Butcher had done it again.

This fish also came willingly into deep water but soon began to show an alarming tendency to go further afield so that we were soon in fast pursuit Angus driving the boat as only he can. In heavier water we should not have had that fish because he persistently worked so far away from the boat that I had no reserve of line. At moments when he did come slightly nearer I regained line by hand drawing, clamping a finger of the rod hand on it and then getting the slack on to the reel. I dare not allow the reel's vibration to travel directly down the line for fear of stirring the fish into a burst of running. He took longer than his predecessor and when he did come to the net I thought we were fortunate to get him since he had been on far too long. This was also a fresh fish – a little heavier than the first and when he was tapped on the head and the fly passed back to me, we had a smoke and a think.

The loch still glittered in the sun and my face burned with the reflected glare. Whilst it was true that with fish moving up the loch as they obviously were, we would come across one almost anywhere, it is a big sheet of water and not all of it fishing ground. We were in a trout fishing area, the day having been so planned, and there remained of true salmon lies thereabouts, only one as yet unfished. It lies off a low peaty headland where shallow water covers a stony flat reaching out into the loch. The bottom is not as clean as I like to see it – hard patches alternating with soft ground and I have never had much faith in it which is perhaps why I have never had any success there. Angus used all his very considerable boat-craft in fishing it but we failed to move anything and having covered it moved back towards the bay entrance where we had the first fish. Another fish *might* have occupied the lie – if lie it was – but if so he did not want the Butcher or the Silver Doctor on the tail.

We ended our active fishing pulling slowly and diagonally across the bay to the boat landing while I cast a long line, more from habit than from hope, at right angles to our course and allowed the flies to swing round towards our wake. On some lakes, in the spring fishing, this will raise more fish than conventional casting. It is not so every day and I have a feeling that there is some condition of weather, wind and wave that favours the method. I have never been able to find enough constants to identify the condition and know only that it works better on a rough day than on a quiet one – but then, so do other methods.

Anyhow, that is what I was doing when Angus shipped his oars and allowed the boat to glide over the shallows to the beach. And there, in eighteen inches of crystal clear calm water over sand as flat and featurelesss as an area of concrete, I hooked a third fish. The boat grounded and the fish jumped. We saw he was small – I concluded I could play him from the stern with the boat where she was, but Angus pushed off ten yards to give us mobility. That little fish gave me much more trouble than either of the others. He was everywhere in short dashes and lashing jumps, stimulated probably by the very shallowness of the water. It was good clean fun until the reel retaining band, which must have been working loose for some time, slipped off the plate and the reel fell at my feet. A return from there would have been simple but the reel disappeared from sight where floorboard failed to meet stern frame. It continued to revolve, however, when I needed more line and with care to stand still so that the yards of loose line came up from the boards unchecked and by guiding payout from a light grasp on the line, the fish was played. It was by no means as difficult as it might appear although I would not care much for it with a big fish.

This one was an elegant grilse of four pounds so that we ended this day of surprises with two salmon and a grilse and twenty-one brown trout. It kept us talking all of the way back up the other loch and over the hill to home.

I was happy that evening to enter in the estate fishing records the first capture of salmon in June and doing so under the fourth of the month was prompted by some quirk of fancy to look, several volumes back, at results of other anglers on other fourths in earlier years.

The entry for the fourth of August 1914 read: 'War declared on Germany. Wind went south in afternoon.'

IX

Running Salmon

THE belief that salmon will not take whilst they are running is so generally held that it amounts almost to one of the axioms of angling. I do not share it. Like so many other fixed ideas, it has no unshakeable ground for its fixation because it is only true in part. The extent of one's belief is directly related to width of experience – not necessarily to its length. If I had fished all the time on any one of a dozen rivers I could name, I think I should believe it too. As it happens, I know other rivers where fish *will* take whilst running, though I will admit that the process of catching them contains a considerable element of opportunism and, what is more, requires an intimate knowledge of the water at the critical height which induces the running mood in the fish.

Since so many inactive days can result from a belief in the futility of fishing when it is said 'they're running', it seems worthwhile to look fairly closely into the conditions which so often lead the fisherman, if fishing at all, to do so without confidence. Lack of confidence is bad: I am not sure it is not the greatest handicap of all. All the old precepts, like 'keep the fly in the water' go some way to make up for it but it is the heart and feeling and underlying strong hope that goes with covering the water that counts most.

What makes salmon run? It is obvious that there are two kinds of runs, the run proper which is the entry into the river of fish fresh from the sea whatever the season, and the subsequent upstream movement of those fish from pool to pool and beat to beat. Both types of run are most readily triggered off by a flush of well-oxygenated fresh water. It may be a big spate or a rise in level of a few inches only – it may be snow water or relatively warm and the salmon's response is conditioned by the season, the time the fish have been waiting for the water and its temperature and the spate's duration. If it is a long-lasting, moderate or big spate, late-running fish especially of the summer class, may easily outrun fish of the spring runs so that, and this is almost always the case, one finds stale spring fish in lower or middle reaches and fresh fish high up the river. Thus the stock in any given beat can be and usually is, mixed. A little watching bears this out and most fishermen know it. But there is more in it than just the appreciation that the stale ones are the least likely to take and the fresh fish the most promising. The fish behave differently in their manner of running.

To look at a typical running water we have to assume that season, water temperature and its origin and oxygen content suit the fish. Salmon begin to be restless, particularly after a period of low water at some time, it may be an hour, in advance of any indication of water level change that is perceptible to human eye and may then take while the water rises an inch or so. Then they go 'off' as water continues to rise and fish move out of their lies and from pool to pool. If the river is rising slowly this upstream movement may proceed and the fish remain unsettled and moving slowly. If they can be called 'runners' at this stage, then runners *are* hard to catch. Then comes the main weight of debris-laden water and generally speaking at this stage in most waters, fish do not travel far. If you could see into the murk or the turmoil of water you would find them singly or in small bunches, tucked into quiet places, along normally dry gravel margins and often very close to the bank.

The great mass of fish move upriver sometimes for many miles, when the peak of the flood has passed but just before the river has reached the stage we call 'fining down' – clearing and falling but still heavy and fast. This is 'running water' and it is in this state when we see so many forward upstream lunging fish that are very obvious runners. I know rivers where they simply will not take at this stage, as I have already said, and many more where they will.

I do not know why rivers differ in this way – perhaps they only appear to differ and perhaps what differs is my knowledge of the rivers because I have noticed that those which I find unresponsive are those I know least well, whilst those from which I can take runners, I know intimately. And this latter makes my point about opportunism plus knowledge of the water.

Two examples will illustrate. One is a river in Scandinavia just south of the Arctic Circle where for some years I shared a beat with a Norwegian friend. We had ample opportunity here to fish for running fish but to my knowledge never killed one. These fish *had* to settle before they would take because, I think, there was little or no rock in the river – the pools were floored smoothly with gravel so that the only shelter was in the deep body of each pool. Running fish entered over the tail stickle and without pause went straight to the deep below the head stickle. Some dropped back later but they had, in that phase ceased to run. I recall very clearly the lowest pool on the beat – three hundred yards long, smooth and even and gliding away at the tail to the head of the rapid below. During a run, hundreds of fish would enter this glide in the course of a day, some visible in the clear green water and a large number plunging or otherwise showing as soon as they entered the pool. One fished always over fish in such conditions but neither here nor in other pools would they take. So I say that in this river the runners will not take but I do think that they would take if one could find them resting even momentarily in fishable water. But there is nowhere for them to rest except in the shelter of deep water unsuitable for the fly.

My second example is a Devon river on which I have fished one beat for twenty years, studying and coming to know its moods and more important still, every boulder, every ridge of rock and every quirk of the current in all of its pools. In this river running fish *will* take but not in the places where they take at normal or best fishing height. In a spate salmon will be in strange small corners which a stranger would pass by. Examples are too numerous for detailed description here but, for instance, in the middle section of a long, gravelly shallow there is, under one bank, a short undercut where a diagonally thrust local current has gouged away gravel and clay and provided an extra foot of depth. The gravel is in equilibrium now and current no longer scours but runs across on the surface. The quiet water underneath always holds a resting fish in a falling spate. On the other bank, higher upstream, is a slight depression in the gravel, almost hidden in trailing briars and waterside herbage. Here similarly fish will lie to rest having forced their way through long rough shallows, though it is as bare as a board and just as fishless when the river falls. In the next pool above are two angular stones, embedded in gravel and a foot apart. They project nine or ten inches and between them is a flat stone. Something in the flow between the projecting and over the flat one provides an ease which is attractive to fish and this lie is always tenanted by one fish at a certain river height. It is five yards from the bank – just two relatively small stones like many thousands of others in a stony river. You would pass it by. All three of these places will provide a 'take' from running fish though they never settle there; they may be there for ten minutes or half-an-hour or half-a-day, no more. They are resting places only and if you can find such places – it takes years but imagination and natural divination will help – you can catch running fish.

There is one way in which the long period of learning about the physical character of a river bed can be shortened and that is by weed-cutting in low water. Granted all rivers do not grow weed but in my case I combine a useful job – often waist deep in low July or August water – with study of the pools and runs. I can see no reason why the exercise, in the absence of weed, cannot take place simply for purposes of exploration. Disturbance of the water, in drought conditions, can hardly hurt fishing. And if you go cautiously you will be surprised how many salmon there are that were quite unsuspected and, more important, how many potential lies there are. Thereafter you can fish with a contour map of the pool floor in your mind's eye and coupling this with the general knowledge of the run of the water at all heights, you can present the fly to cover each lie in the way that seems to suit it best. But the lies – for each level of the river – must be known and distinguished as long-stay lies where fish are usually found lying for days or weeks, from temporary resting lies from which one may winkle out a runner. Intimate study of a river or a section of it is a fascinating game and a rewarding one.

I have seen a friend, new to the river, fish my best pool thoroughly and competently while I watched from the opposite bank unable to direct him because the voice of the river drowned my shouts. He missed one small piece – a slight ease between two currents only about a yard or so in each dimension and from my higher bank I could see a fish in it poised in mid-water, active and ready to take. I went upstream to a footbridge, crossed over and came down to the pool. My friend refused to fish the pool again saying, 'You show me how you would fish it.' I did, and that salmon, a neat little seven-pounder took at once and was landed. First you had to know that fish took precisely in that yard or so, and second that you fished one of the currents and then led the fly over to hang in the ease by putting the rod over towards the other current and your bank.

Lastly, on a complicated subject which contains more than its quota of exceptions and reservations, I must say that I know two rivers in the Scottish north-west where fish can be caught while actually on the move upstream. Resting fish among them will take, of course, but fish will also snatch at the fly while moving from pool to pool! There can be no doubt of this. In one case water is always clear and the running fish can be seen to turn aside and take; in the other, fish, though not visible once they have entered the pool, will take while going up-stream through the sea-pool and I have several times watched a fish come in over the dam, put the fly in the path of his bow-wave and caught him. In these two cases the fish were of course absolutely fresh.

X

Saltwater Salmon, and Seatrout

QUESTING among anglers for aspects of fishing which are little enough known or written about to excite interest today, I found in the salmon and seatrout persuasion a very live and unsatisfied curiosity about salt water fishing. Orkney and Shetland voe fishing for seatrout features fairly well in the literature of the sport and most seatrout anglers know too that seatrout can be caught in the vicinity of river-mouths round the coast and specially in the West Highlands. But I feel that the scale on which this type of fishing can usefully be carried on is not understood as it ought to be, and salt water fishing for salmon immeasureably less so. There would be much more salmon fishing in the sea were it not for the widespread belief that salmon will not take in salt water, and for the fact that not all river mouth environments are suitable: the point is not that the fishing must be conducted at a river mouth but that there must be a fish-attracting inflow of fresh water to perpetuate the shoaling habit in fish moving inshore from outer waters.

I could take you to places where you could throw a fly among shoals of salmon at the choicest states of tides without moving a fish so that you would say 'they' were right and salmon do *not* take in salt water. And I could take you to others – selecting tide and time and wind conditions of course, where your first cast among the fish would produce a fierce pull and a take. You would say, 'Why haven't I heard about this before now?' It is not however quite as simple as, in order to make my point, I have made it out to be. Whether one seeks sea-trout or salmon the subject is one that requires study. All worthwhile fishing methods require study of course, but this one does not involve the mechanics of fishing nor, within small limits, special equipment. It does require meticulous study of the habits of the fish, and of the local conditions, especially the tides. If you are fishing river or loch you take great care and appreciable time to investigate to the point of thorough understanding, the run of the river bed, the effect on current and therefore on holding places, of bends and spurs of rock, gravel bars and so on. On the loch you study the bottom contours, the trend of the hidden currents and much else besides.

In salt water fishing, for salmon especially, local topography will require this same degree of attention, though the identification of holding places has no significance because you are looking for and will be fishing to, travelling fish. It is the finding of taking places where salmon will come to fly or a spun bait that is vital. And the difference between some shore-line point where salmon can be taken from passing shoals and those where the fish ignore anything you care to show them embraces a whole unique range of frustrations. Substitute for the shore point, if you wish, a boat well handled and skilfully held in position on an interception line, but you will, if my own experience is any guide, only succeed from a boat if there is adequate wind and wave. I have found seatrout markedly less nervous of a boat than salmon which is odd. I regard seatrout in fresh water as the shyest fish but their shyness is something they acquire progressively as they leave their salt water environment behind them. In the sea they are easier to fish from shore because they will take in a vastly greater variety of conditions than will salmon, they are less defined in their travelling routes, and they are somewhat less critical of tide height. They will also take in a greater range of depths though tending to prefer shallow or medium depth water and they will take in confined spaces among rafts of weed and very close inshore. The seatrout is at home in the estuary and nearby waters, the salmon is an ocean feeder and whilst he too is, of course, a fish of two environments there is a greater difference between his ocean life and the inshore and river portion of it than there is in the similarly dual life of his smaller cousin. To the difference between their sea lives I attribute the difference in their reaction to any fly or bait presented to them and in this context it is worth remembering that while seatrout carry their feeding habit into their estuary and river life, the salmon does not. He ceases to feed at some stage unknown to us but certainly before he enters shallow water.

This is the heart of the matter, in my view. If you are going to catch salmon in the sea you have to fish to them in deep water; they will not take where you can take seatrout easily – in shallow estuaries. I am not contending that 'deep' inshore water compares with the real deeps of the open sea feeding grounds, of course, but simply that arrival in coastal water seems finally to inhibit the feeding urge. I find it very hard to answer the question 'What do you call deep water?' because for one thing, moving fish are travelling with the tide in a constantly changing depth. Let us say that you will not catch salmon in any area which is exposed or very shallow at low tide and that I personally have had little success, at any point where, at half-flood there has been less than say, twenty feet of water. Here is where our study of local shore contours comes in and here is where the fortunate fisherman is he who can fish for incoming salmon where the river they seek enters at the head of a steep-sided and therefore (usually) deep sea loch. Many on the western seaboard of the Scottish and Irish mainlands are like this but the islands generally are less suitable because of the machair or other low ground fringing the Atlantic coast. They have other attributes however – in a few cases some salmon fishing in brackish enclosed waters or coastal lochs to which the tide can penetrate.

Granted that you have access to a deep-water channel traversed by companies of fish making for their river immediately prior to their run dates, the preliminary to an attack on them has to be observation from some vantage point at various states of tide. I would begin by being in position at the first of the flood to see what number of fish came in during the whole of the tide and what route they followed, that is to say, how far off shore, how close-in on points – whether they pause to cruise round the mouth of any small stream and so on. Next to be decided is whether the fish are, in fact, running the river. You probably know this by other means – the local bush telegraph for example. If they are not going up, they will come down the estuary and go seaward again with the ebb tide and that should, therefore, be awaited. The best of the salt water fishing comes, naturally, when salmon are collecting about river mouths but are unable to run for lack of water; they then move in and out with the tides and will do this until the river rises and draws them up and away. Meanwhile their tidal movements will be on the surface and you get repeated chances to show them the fly. If, on the other hand, the adjacent river is big and pushing fresh water far down the loch, there will be decidedly less surface shoaling though possibly more individual plunging. Generally the fish are travelling faster, deeper and unseen.

In short, therefore, we need a period (which at the time of maximum run need only be short to permit an accumulation of salmon) of warm, dry if not fine weather. I have found that weather really matters little so long as the river is low but wind from a warm quarter is necessary. It need not be strong so long as it will raise a brisk ripple. Salmon *will* come to a fly in a calm but they will not take it. The leaders of a group, company or shoal will turn aside to follow a fly or bait put across their path but without a ripple will turn away just as you think they are about to take. They are interested but not enough. This is the maddening part of sea fishing for salmon – so often they show interest and so seldom take. The answer has already been given – fish over deep water and in a good breeze.

The critical features are simply location and weather conditions; the actual fishing is simple. Rod, reel and line can be whatever you use on the river but do not use your best outfit; I find that however carefully you wash reel and line in fresh running water after fishing in the sea, you simply cannot wholly offset the corrosive effect of the salt water on metal components and the reel is particularly susceptible. So I use my second best outfit and wash very thoroughly as soon after tidal fishing as possible.

Brightness seems to be the basic requirement in either fly or bait. Flies of the general air of the Silver Doctor, Dusty Miller or Dunkeld category have served me very well and in a rough wave I would go up to fours in size. Alternatively a three-hook lure, say two inches long Dunkeld type in dressing will take very well. The cast can be 8/5 or 9/5 or of eight or ten pounds breaking strain. I do not feel that small flies and light casts enter into this fishing at all because in the conditions which demand them – shallow, calm or lightly ruffled water – salmon will not take although seatrout will. Spinning, on balance, will take more fish than fly fishing and I like a multiplier with ten-pound line and wire or nylon trace with a blue and silver one-and-a-half inch Devon as well as anything. I do not think bait size is critical though you should err on the small side; the water is commonly very clear. A small spoon, gold inside and hammered silver outside will kill. Rod, much as I personally like the seven- or eight-foot single-handed type, is best rather longer. Sea-caught salmon, especially grilse or small fish, are excessively soft in the mouth and with a short rod have to be held much too hard as they are brought into the net.

So often the last plunge into the fringe of wrack along the shore requires the 'holding off' that only the nine- or ten-foot rod can provide. And it is small consolation, having hooked a fish out of a passing shoal after a long wait, to lose it at the net and recover only a fly hook or triangle ornamented with a small strip of mouth tissue or an inch of maxillary.

Before I turn to seatrout, two last points and first the

tide. Half-flood seems to bring most fish but this can vary with local conditions. In most places it pinpoints the maximum willingness to take whereas the ebb, in general, finds the fish in different mood. It is almost as though, having failed in their natural objective of running the river they had lost heart and interest and on their way back to deep water have much less regard for fly or bait. I do not say they have none, and in some places the local fishermen favour the ebb.

My second point concerns sea-pools of which the variety is infinite. If they are deep and wholly salt at full tide they are worth fishing though fly size may have to be adjusted downwards and cast weight proportionately. If, however, as so often happens they are shallow and sandy or pebbly the chances with salmon in my experience are small and you should turn your attention to sea-trout. Salmon, however, have no book of rules or, if they have, they often ignore it and an odd fish can be taken from the unlikeliest places at times. An undercut rock in a sea-pool will often be occupied by a fish coming in with the tide and such a fish may not go back with the others as the tide falls. Given a wind on the pool, you may catch that fish later irrespective of pool height. But one cannot call this salt water fishing: indeed at low tide the pool will not be even brackish. There are a few places in the Hebrides where lochs close to the sea are invaded by the tide and whilst of most it is said that neither salmon nor seatrout will take when the loch is brackish, there are exceptions. For my part I have found salmon terribly stiff in mixed water conditions but seatrout very much better to the extent that they will take in river mouths where salt and fresh water mingle in distorted spirals over the sand.

But true sea fishing for salmon must be done out where the sand is hidden by sheer depth; here fishing to the leaders of questing shoals as they go up-tide tightly packed, an individual here and there plunging ahead, just beyond the floating weed, can be very exciting indeed. The fish are unbelievably strong and active and 'hands' are needed; hard holding will provoke violent acrobatics, uncontrollable runs and perhaps pulled-out hooks. Too light a hand will result in fish weeded and lost.

Sea-caught seatrout in play, and size for size, are similar. Mouths are even softer than those of the salmon and the fish have a much greater tendency to dive into weed. The spinning rod is usually too stiffly-actioned and rather than spin for these fish I use the two- or three-hooked lure in the salt water, and flies – preferably on the small side – in sea-pool and tidal lengths of the river. Seatrout, however, differ from salmon in one important respect and that is the variety of salt water conditions and localities on which they can be taken. Brief reference has already been made to the voe fishing of the northern isles. The technique can be used equally well wherever seatrout feed in sea lochs along the fringe of the weed. Fly rod and tackle are used to cast standard pattern flies – Butcher, Blue and Silver, Alexandra and others in the attractor class and the size can be what you would normally fish in neighbouring freshwater lochs. The worm on Pennell or Stewart tackle, cast like a fly – but more smoothly and carefully in the back cast! – and worked rather slowly, can be deadly, though I do not think any superiority over the fly outweighs its disadvantages. The worms do not live long in the salt water and even with toughened baits there is the nuisance of frequent re-baitings.

In the open estuary seatrout will take flies of the type indicated. Lures made up from strips of teal with silver bodies and blue hackles, those with peacock strips for wings, silver bodies and red hackles, and indeed almost anything with flash and bright colours. The weakness of the lure is its fragility. Elaborate mixed 'wings' are unnecessary and soon fray to nothing – the simpler a lure is the better. As with salmon some ripple on the water is a great help. Seatrout will cruise in small lots or big shoals in very shallow water which is often crystal clear. In calm conditions these fish can be shy but they will take under any reasonably concealing ripple. The shoals are easily seen and if the fishing point has been carefully chosen it can be possible to cast to each of them as, ruffling the surface in a fairly compact mass, pushing the collective bow-wave ahead of them, they come abreast. Usually the majority of such passing groups will be within reach – they seem to hug the edge of the weed on wrack-fringed shores though much depends on shore declivity. Again, I prefer the locale where the loch is narrow and shores steep rather than the estuary where water is thin on sandy channels. Incidentally this sandy channel fishing, common and very productive in some of the Western Isles, demands the maximum local knowledge of tides and quicksands. It is emphatically not a game to be taken lightly.

Conditions for seatrout fishing in sea-pools vary enormously because of the variety of the so-called sea-pools themselves. So very often the lowest pool in the river is invaded by the tide and this, properly the sea-pool, is shallow. The sea throws up a bank of gravel where the fresh water goes into the salt and this holds up a stony reach of thin water in which seatrout (other than finnock, herling, whitling or what you will) do not stay, going back again with the tide. I know one such pool which after each summer tide is literally floored with finnock and never a good seatrout among them. Other pools providing depth and cover, even at full ebb, are different propositions entirely and should be fished at any state of tide. Some fish will stay and I would fish for them as in a normal river-pool with normal flies and lighter tackle than I would use for the open water of the estuary or down along the shores of the sea loch. All sea-pools within my experience tend to fish best at dusk and after.

There is another kind of salt water fishing for seatrout which in some ways is the most interesting and exacting of all. Unfortunately the number of localities suitable is very limited because proper conditions can only arise where an enclosed area of seawater forming feeding ground for the trout is filled and emptied by the tides through a narrow channel. There are a few such places in the Outer Hebrides and some on the north Irish coast. It is important at this stage to understand that broadly the fishing I have been describing thus far has been based on seatrout within hours or days or weeks of their summer or early autumn running times, but that we are now considering what is mainly a spring and early summer fishing – in short a fishing for seatrout while they are living their normal feeding and growing life in their sea environment. Most of the fish one meets will, it is true, run their rivers in the next two or three months but in the spring they will be mixed in the sense that some will be newly down from their previous river run and are thin, others fat now with three or four months' rich sea feeding behind them or perhaps with a whole year of it. Some slight discrimination in taking spring seatrout from the sea is therefore needed.

The tidal channels on which one should select a fishing point – preferably with local guidance – can be anything from small river widths upwards depending on the size of the enclosed area and the topography of the surroundings. It can be the tide outlet for a whole sea-loch or an expanse of slob land submerged at full tide, or a freshwater loch which is penetrated by the tide in large volume. A mere backing up of fresh water will not suffice because the river which fills your selected channel at flow and ebb must really be a river with swirling current – the kind of tide race that looks dangerous and can easily be just that. In the places I have in mind large numbers of seatrout rove up and down with the tide as it fills the channel and may be found to take on both tides. My experience in Ireland has been, however, that the strongest rush of water comes with the ebb as the basin empties and that it is the outgoing fish that take best. They are fierce fighters in the strong current and a renewed warning about holding too hard is not out of place here because, again, mouths are soft and hooks pull out. Tackle can be heavier than for static-water fishing and the fish are not shy. I would not hesitate to fish a two-inch lure but a spun bait is best of all – a blue and silver Devon of say, one-and-a-half inches for simplicity and a natural sand-eel for choice. Plenty of line is needed because fish and current are strong and the over three pounds class are very liable to make long runs. I use an eight-foot single-handed spinning rod, a multiplying reel with 150 yards of line and seven- or eight-pound nylon or wire. In this fishing, you do not see the fish as in estuary fishing, you fish the stream as you would a river, the only difference being that here the fish are passing and you rely on putting the bait across their paths; in the river they are, for the most part, in their lies and stationary.

These spring seatrout from the rivers of the sea fight to the last and any carelessness will lose many at the net. The final dive for the weed as the net is extended is especially dangerous. Again, though it is true of playing all fish everywhere or at any time, 'hands' are an enormous asset.

XI

Salmon at Dusk

WHILE it is a commonplace in the writings of our sport to say, and truthfully, that fishing takes one into delightful places there are some places which, of their very atmosphere, provide a special lift to the spirit and there is one lough in my dearly-loved Donegal which has this quality above all others. There are other loughs which to me yield more salmon but less joy in being there and in fishing.

No majestic mountains tumble screes to this lough; it lies on low ground chequered by small green fields and gorse-topped banks and along the flat stony shores thickets of bog-myrtle – russet-flowered in the spring. Columba, the turbulent priest-prince who became the gentlest of saints, walked here and here much earlier Neolithic man built his crannogs and the secret shallow causeways to them where today lurk the best of the lough's trout. Beyond the run of the cultivated ground in the north-west are the tops, softly azure in sun and profoundly blue in shadow, of the high hills. And at one end of the lough, between and around inflowing and outflowing rivers, are acres of reeds sere and golden at the time of the spring fishing – a green waving jungle in summer. Here are grebes and ducks and otters and sanctuary for any salmon with the wit to make a long, strong run into the fastness. The reeds grow, of course, in the shallows and all along their outer limits fish lie thickly for perhaps two hundred yards out into the lough. But before they do so, entering fish travelling in companies after their eight-mile run from the tide, scatter round the unreeded shores and for a time – hours or days – may be found almost anywhere within half-a-mile of the outflowing river by which they entered. And hereby hangs this particular tale.

It began about ten o'clock on an April morning when with my gillie John, an old friend, I pushed off from the river mooring and headed out through the reed-walled channel to the lake. Inside the wind-tossed reeds the waves were no more than an even surge; outside on the open lough there was a big lively wave. Wind, unhappily, was due east and the light was uniformly grey and poor though cloud was high. It was cold too, and all things considered it was a singularly unpromising day. It was the more disappointing because the lake was falling quickly after a spate that had, we knew, brought large numbers of fish, the main spring run. They had had two days to settle and with any kind of luck at all with conditions, we were justified in expecting a good day. John pulled fairly close in on the eastern shore and we began a long drift across the reeds into the western bay and the lee shore. Fish plunged here and there and I had one soft ineffective rise as we passed the mouth of the incoming river. Back upwind we had another drift across slightly deeper water. It was admittedly early in the day to expect fish to move in April but the big stock should have provided an odd bolder fish. It did not. I began to look round and further afield. I have a note in my diary which shows that at 12.20 p.m.

I looked up and saw an eagle swinging slowly very high over our heads – the only eagle I have seen in all my Irish years.

At one o'clock, we beached the boat on the eastern shore and lunched out of the wind under the thick gnarled gorse a few feet higher than the lough, the whole fishing area in sight. 'If only' – we said – 'we had a west wind!', but the day looked, if anything, dourer than ever and in fact we fished the afternoon out without seeing the slightest sign of an interested fish. Not surprisingly, the usual afternoon move of trout failed to materialize. In that bitter wind there was no hatch of fly and the overall conditions keeping the trout down were of course affecting the salmon. I always believe that if air conditions will stimulate a rise of trout there is a chance of a salmon. I have seen it happen so very often.

The day so far, however, was not without a certain reward. Three successive drifts ended close in to the shore on the west – with the prevailing wind a lee shore – on which the rollers broke in foam on the stones. Each time we approached the shore we saw plunging fish and they seemed to be concentrated among the submerged ruins of an old shieling – the temporary resting-place, clearly, of a shoal of fresh fish. I fished down to them and among them once without moving a fin and thereafter stayed away lest we should scatter them by putting the boat over such a shallow and limited lie.

At five o'clock I went ashore to sit by the fire and drink hot tea in John's little house. Here soon I was joined by an old friend from the village up the road. Like me, he was haunted by the prospect of a lough full of fresh fish and, despite the weather – now steadily growing worse – and refired by his enthusiasm, I decided to try again. My diary records that we fished, he in the bow and me in the stern, from seven till nine.

We worked across the reeds first, the wave much bigger now and the light going rapidly. Because of both wave and the overcast I went up two sizes in the flies, with a Claret on the dropper and a Thunder and Lightning on the tail, both fours. Down in the corner where the reeds grew thinner as they met the stony west shore I had a good rise, a solid take, and the first fish of the day was on. Hustled a little because we were anxious to get at what John described as 'that nest o' fish' among the old walls, this one, a lovely fresh hen of twelve pounds, came to net fairly quickly.

The light became less and less and very soon we were fishing in a kind of weird dusk lit by dirty yellow gleams from the west where angry blue-black cloud masses parted fleetingly to admit what remained of sunset. Along the shore where we had earlier marked the fish, wind and wave were too heavy to allow us to drift broadside and with the bows half into wind, I in the stern, had the best of it. How many fish we rose I do not know. They rose like trout with quick slashes at the fly and the lurching boat made it difficult at times to leave it with them.

My friend had first blood – a ten-pound fish that led us away up-wind before we had him. At the net he did a complicated lash and roll and left the cast hopelessly entangled and possibly weakened. A new one went on while we went back on the drift. Both fishing again we had an immediate demonstration of how clearly a fish can see the fly even in bad light. My friend's tail fly was six inches in the air and about to hit the water when a fish poked his head out of the side of a wave and took it while it was still in the air. We all saw it and the twisting-away snatch that followed. I would not care to say we heard the 'snick' as the fly parted from the cast but the picture was so vivid before us that we imagined we did. Then it was my turn with another up-wind quickly-beaten runner and again of ten pounds. I knew that we should lose one of these active fish sooner or later because strong as John is, it was impossible to move the boat head on in that sea at anything like the fish's speed and one saw backing going and going. . . . In that lough too, fish are exceptionally strong and with normal tackle it is quite impracticable to try to stop a fresh fish determined to go. In a light breeze, with easy manœuvrability of the boat, such fish can be turned with sidestrain but not in heavy water. Here we were, wallowing about in near darkness and it seemed that every fish hooked was crazy to get to the middle of the lough.

The next fish my friend hooked was assuredly crazy, running round and round the boat with a peculiar wriggling motion on the surface for all the world like a kelt, though he was not that but a very fresh eleven-pound cock hooked firmly in the tongue. At the first cast, after boating the wriggling one, he had another and this one for a change ran inshore, lashed like a fury in very shallow water and smashed the cast. There was a great deal of dark muttering up in the bow.

It was dark now, as dark, John said, 'as the inside of a cow' and I think we were all becoming slightly insensitive – buffeted by wind and wave and perhaps a little too exhilarated to fish carefully. I know that I missed two or three fish that in daylight I think I would have hooked – the quick splashy rises were difficult to see and worse still to time in the rough water. All the fish came to the Thunder which of course suited the darkness and with his broken cast my companion had lost the last big fly of this pattern he had. I gave him one which later, in the light, turned out to be a Black Dose. He had no more rises but I had a third, rather smaller, fish before I too was broken.

This last fish fulfilled my earlier prediction. I hooked him inside the ruined walls and he shot out past the boat like the proverbial greased lightning. Straight out into deep water he went; the reel screamed and went on screaming while John, nearly beaten now, struggled to force the boat upwind. We went slowly after the fish and for a time the reel was silent and I began to hope. Somewhere about ninety yards ahead in the tossing blackness in the eye of the wind there was a very strong fish. The rod was bent with all the strain the tackle would take and my finger on the line when the rod point was pulled down hard and I *had* to let more line go. Inevitably after that I had to try once more to stop him to save the last of the line and this time the cast gave between dropper and tail and he was gone. I suddenly realized how cold it was.

Half-erect in the lurching boat, we put John in the stern with his feet among the fish, and side by side amidships we fishermen pulled home. I have taken odd fish just before dusk but only odd ones. A take such as I have described, yielding five fish in a bare two hours, much of it in total darkness, is an event.

XII

Seatrout in the Night

If seatrout are less mysterious than salmon in their comings and goings and in some aspects of their life history they are further beyond comprehension in other ways. A switch of wind, weather or light, for example, will oftener than not bring stiff salmon 'on' but seatrout are not so easily to be moved by direct influence. Their waywardness is much more deeply rooted.

It shows itself in their response to night fishing, which varies in success not only with weather conditions irrespective of locality but from one district or river system to another. Overall, their caprice will increase with lengthening sojourn in fresh water. But comparing fresh run fish in one area with those in another part of the country and, similarly, stale fish with other stale fish one wonders – at least I do – whether the differing characteristics of local races may not be at work to produce these inconsistencies of behaviour. I am sustained in my ideas on this subject by the knowledge that fishery investigations have established beyond doubt marked differences in the growth-rates and ages and in a number of staging points in their life histories of fish of different watersheds – even where such watersheds are closely adjacent. More than all this, the fish have been found to differ *in habit*. My own observation only emphasizes the variation in habit. The differences in appearance which derive from life-history, feeding, environment and a host of other factors and provide the inherited characteristics of any river's race are sometimes very plain to be seen even by the uninformed. I have heard non-fishing house guests, gathered round the fish slab when the fishermen return from forays to several rivers at the end of the day, exclaim in surprise at the range of shapes, spot frequency and coloration among the fish. Some of us could tell at a glance where each 'type' of seatrout came from.

The variation in habit is a different matter; that is obvious only to the fisherman for he encounters it when he finds that the behaviour pattern in the fish – exploited successfully in one river or loch – is useless in another and perhaps not far away. By comparison, the variations in salmon behaviour are minor. The constants in seatrout reactions are very few, which alone would make their beguiling the absorbing passion it becomes with some of us.

Local race influence or not, the reaction of the fish to the presentation of the fly in one district, place or river as compared with that of another is, it seems to me, most sharply marked in night fishing. On the basis of twenty-five years of seatrout fishing at all hours in the south-west of England and over fifty years in pursuit of the fish in Scotland and Ireland, I have come to know that night fishing is not the universal remedy for poor results that so many believe it to be. Indeed in many areas it is simply not worth the time spent on it – in others it may offer the only hope of a brace or two, or a bag of seatrout. High water conditions are excluded of course and I write only of low water and summer weather in which in daytime seatrout can be as elusive as only they know how to be.

In the West Country with rivers low and seatrout staler day by day you will, if you have the capacity to work your water with all the care required to walk on eggs, catch some seatrout on dry fly. With wet fly you may pick an odd 'schooler', probably a whitling, from the runs and nothing else. But on the right night with, admittedly, some egg-walking capability, you may take some rather nice bags of seatrout including the odd big fish. Find similar conditions of weather and water in the Hebrides, parts of the mainland and parts of Ireland, and your wet fly (and your dry fly) will take fish in the day and sometimes all day – but you can go home about dusk because it is finished. I am not prepared to say that all Hebridean rivers and lochs are like this – all that I know are. I have fished the most beautiful pool tails at dusk or later and again *much* later and lochs too – both full of seatrout – and failed. Someone said, 'Success knocketh at the door but failure walketh right in.' In night fishing in the Islands, in my experience failure walks right in. It does the same thing in many rivers and most loughs of the West and North-west of Ireland. Across the Irish Sea on the other hand, many Welsh rivers are renowned for the quality of their night fishing though they can be as unimpressive as those of Devon and farther west, in the daylight. In a general sense it can be said that after-dark fishing for seatrout is not a method for loughs or lochs: in most of my fairly wide acquaintance fishing ceases at dusk – in some cases long before – although the inflowing or outgoing river may fish well after batflight. I know a very few lochs in Scotland that fish at night. It is significant that in these few places the fish are taken on selected shallows or along the margins; it means, in my view, that the bulk of the fish are bottom feeding if they are feeding at all after dark, and are not alert for surface food. So, as a rule, night fishing is river fishing.

In those areas where night fishing is not successful one never hears it suggested or mentioned and activity on the river after dark may indeed be viewed with a degree of suspicion in some districts. Where it *is* successful it is the accepted process of angling and day-time fishing in a low river full of shy and dour seatrout regarded not only as a mild form of insanity but also as damaging to everyone's prospects by making fish even shyer. Seatrout certainly can be flogged into extreme dourness very quickly and daytime over-fishing will injure late evening and night fishing results.

The darkened river is another world. Its sights are hidden, its scents and voices intensified. Intimate know-ledge of the water is an enormous help: a periodic break in the sound of a shallow speaks of even flow broken by a big stone and, knowing the river, you know exactly where that stone is and what it looks like. The blur under the far bank, faintly lighter than its background, is the projecting tussock of bleached coarse grass where the otter sits, and down in the tail glide there is a place where you must feed an extra yard of line if the drifting flies are not to skate. You know when one of those hard-fighting three-pounders goes racing away into the black cavern of overhanging trees, what he can do under there and how far you can let him go because the bed of the river is familiar to you – and so on. That is knowing your river.

Seatrout will not be out in smooth thin water in

summer sunlight once they have lost that first freshness, but they will spread out over such places in darkness. As dusk deepens they drop back out of deep water, emerging from cover into the tails of pools, the smooth glides and the flats of even depth. Some may go upstream into broken water but the majority go down and a broad evenly-flowing pool tail or a long flat which shallows slowly as it goes down are places to choose, particularly if bank configuration is such as to hide you in the shadow. The ideal – seldom found I admit – is the place where high ground or trees prevent your appearance on the skyline but are far enough behind you to give an unobstructed back cast. Trouble-free fishing in the dark requires nice length judgment in the forward cast and this is quite inconsistent with the need to make steeple or switch casts or to execute unorthodox evolutions in the path of the rod. A very simple fishing style, quite unhurried, is best. With two widely-spaced flies and a calm approach, there is no real reason why inability to see what is happening should take the pleasure out of fishing as it does for many. Quiet unhurried movements are vital anyhow; seatrout are sensitive to vibration, waves from wading, lights and even the glow of a cigarette. Lights are bad for two reasons – they frighten fish back into deep water and they also hinder the fisherman's night vision – the vision he will have developed if he has been watching the water – without laying a line on it of course – since before the bats appeared. I always endeavour to unhook fish or change a cast without use of a torch – the fish I deal with by touch, the cast partly by touch and partly by sight against the sky.

Of all weather conditions, I think on balance I prefer

it to be warm and rather dark when I fish at night. Moonlight is a mixed blessing, making it easier to move about without blundering into fallen branches, ditches, wire and other hazards, but making some pools unfishable because of shadow – angler's shadow. On flats a warm light breeze is an advantage. Rain makes very little difference except that the thickly-coloured sky which produces it may give the night that woolly black gloom which can be baffling even when you know the river bank well. On one such night although I had been out since dusk and my eyes were well-conditioned to the dark I fell over a resting cow in passing from one pool to another and threw the rod clear by instant reflex. Then came the problem of how to find it and the wonder whether the startled cow would trample over it. Finding it required the use of the torch and once it had been used I was helpless without it.

Some anglers I know have a liking for moonlit or 'light' nights and some do well but my best results over the years have been in the degree of darkness which makes a hooked fish very difficult to see until it turns over in coming to the waiting net. Darkness, however, is not simple to define by degree and happily for fishing purposes there can be a reasonable tolerance. There is just one cardinal rule and that is never, having selected the pool or pools to be fished, begin too soon. There is a critical period, about dusk when fish have moved into their night feeding quarters in shallower water than they have occupied during daylight, when they have not settled. If they are fished over at this time they will dash upstream again and though an odd finnock or whitling may take, the bigger fish which are extremely wary in low water, will not take and will not reappear.

In short, that particular pool will be ruined for the evening. It is better to be slow to begin than to start too soon. The chosen pool should be approached by detour if it is open, and under cover if treed or high-banked, at or before the bats fly and stealthy observation kept on it. If there is any gleam on the water the bigger fish can often be seen pushing waves on the quiet surface as they move into position from day lies. Temptation to show them a fly must be resisted until it is dark enough.

The time of a take varies enormously not only from river to river but from month to month through the season and from day to day with the weather, and any attempt at guidance on this is no service to you. After a few nights' fishing and in the absence of any marked weather change, some semblance of a pattern will emerge. Sometimes the fish 'come on' early, moving for perhaps half-an-hour or so and then the river goes dead. On the succeeding night the sequence may be reversed and on certain nights, by all human standards giving good fishing conditions, they do not move at all – unless it is before dawn when most fishermen have gone. Seatrout are as capricious in their taking at night as they are in the day in the choicest of loughs, though when they are really 'on' sport can be superb. The proceedings up to the point of netting any reasonably fresh three- or four-pound seatrout at night, especially in the kind of pool which bristles with tree roots and trailing weed, can be rather lively. For once the scales are weighted in favour of the fish.

Very big fish can sometimes be made the subjects of selective attack though only inside a rather narrow band of circumstances. These require that the fish occupies – and is seen to occupy – a daytime lie in medium-depth water say three to four feet and just above the pool tail. At night that fish, if he is of the five pounds or over class, will not go very far. He may at dusk make a cruise, salmon fashion, round the pool but in nine cases out of ten he will settle for part of the night a little lower downstream than he was in the day; he will not drop back to where the smaller fish are likely to be, just above the draw of the extreme tail. With his day lie well noted the area about and below it can be fished very carefully at night with one fly on 1x or the equivalent in nylon. These big fish know a thing or two, they have been up to spawn several times and the finer one can fish with safety the better. But the class of fish I write about cannot be held on light gear and I think it foolish to try. I have had them up to nine pounds and I have seen one of twelve-and-a-half taken on a big dry fly – fished quite blindly upstream. For my own wet-fly method I like to fish a fly as big as a number six down and across – the fly drifting at current speed and quite unworked. And for these big fish I have found that whilst a silver bodied pattern like Queen of the Water is good – a brown type fly such as Brown Owl or Mallard and Claret is even better and most of my big seatrout at night have been taken on sober brown-winged patterns with bodies of wool and therefore bulkier than the tinselled patterns.

Mention of the fly working at current speed brings up the intriguing question of how the fish take at night and there is extraordinary variation in their manner of taking. Much, of course, must depend on depth of water and speed of current and there will sometimes be a splashy rise or a boil which, catching some gleam of light, will be visible. Usually the fish will be on and felt before the strike, or will have turned away short of the fly. If you are fishing correctly the line should be pretty taut anyhow, short of causing the fly to skate. At another time, in any kind of water, the take will be a hard pull and the fish hooks himself and will be giving the fisherman plenty to think about, in one split second thereafter. Yet there is a further mood in the fish which can tax anyone's skill – that in which they take the fly so gently that all one feels on the line is the smallest touch – as if the fly had momentarily fouled a drifting leaf. Many big fish take this way and it is one reason why, fishing specifically to one big fish, I fish at current speed. The other is that the fish take more readily at this speed, but the prime reason is that only with an unchanging length of line and the fly slowly working its arc across and with the current can the minute touch of one of these extremely cautious takers be felt. There is only one way to convert these mouthings into hooked fish – a lightning fast strike.

Why seatrout will take a fly so gently – indeed daintily – on occasion is puzzling. The only constant I have been able to detect is, I think, that it seems to occur oftener on warm drizzling nights than on any other but this idea – I am not sure enough of it to call it a conclusion – does not lead us anywhere because we cannot know what underwater factor is at work on such nights and absent on others. Certainly a few nights in every season will yield examples of it and there will be more frustrated than outstandingly successful fishers stumbling homewards up the dark lanes from the river.

To summarize this assessment of night fishing I would say. Make certain first that seatrout are *known* to take after dark in the river concerned. Then choose medium-dark nights when the river is low and the air warm, avoiding like the plague those when there is a sharp temperature drop after sunset; a creeping mist will almost certainly appear by and on the river, probably just at fishing time and fish will not take once this happens. Fish slowly on still or very slow-moving water and at current speed elsewhere and strike at every sign of a take whether seen or felt, every pull, touch or visible break in the water.

Flies for the bigger class of seatrout have already been suggested. For general night fishing I use eights or nines and fish two on a three-yard cast. Patterns can be Mallard and Claret, Peter Ross, Blue and Silver, Brown Owl, Invicta, Woodcock and Yellow, Queen of the Water (especially for West Country rivers). There are of course legions of local patterns especially in Wales where many purely local flies are hackled patterns thinly dressed in the manner of sparsely hackled 'spiders' in vogue for upland North of England and South of Scotland trout streams. On moonlight or 'light' nights some quality of flash in the fly may be an advantage and the silver-bodied patterns should be given preference. On darker muggy nights when the sedges are up I would rather fish with fat-bodied dressings like Mallard and Claret and Woodcock and Yellow. A fly which *may* account for a salmon in the dark – although it is only a faint chance once the light has gone, is the Sweep. When salmon come up out of the deep to cruise round in the better-aerated water at the run into the pool, usually just after sunset, there is fair chance with the Sweep fished on the dropper across the fast water. But one must be sure to confine the effort to the top run or the pool will be spoilt for seatrout fishing later. In late summer and early autumn most of these salmon are very stale anyhow but the urge to show them a fly can be very strong when they raise humpy bow-waves as they surge up and down the pool.

The events that stay in a fisherman's memory can be minor ones which possess and retain their own importance because they are parts of the interest of fishing. Fishing is the range of things seen and felt and heard by the river and on the lough, and the place it occupies in the mind is not filled with reflections of extensive killings of fish, the making of gigantic bags. We all like to do well as a rule and very well now and again, but for me, the golden days are those in which modestly successful fishing is enlivened by minor events, the unusual behaviour of a salmon, the clearly-heard whirring of a seatrout's fins as he jumps in a darkened pool, the poise of a dragonfly and the matedness of a pair of tripping sandpipers on the shore.

Night fishing for seatrout – always a source of some incident because of the conditions in which it is carried on – has a store of memories for me and I shall take from it two tales I think worth the telling.

I was fishing, desultorily by day and very keenly at night, on a tributary of the Lune when that tributary, a pleasant little river, was full of fairly new-run seatrout. Water was very low indeed and there were days of blazing sun. I knew the water but not really well, but my friend, a local worthy, had known it all his life and together we made moderate bags of seatrout up to the two-pound class on most nights. We did not have long rows of fish to photograph on the lawn in the mornings but we usually had six, eight or ten fish to tumble into the kitchen sink on our return in the small hours.

On a neighbouring beat there was a certain famous pool called 'The Wheel' and thither, with the requisite leave, we went one night – my friend most anxious that I should sample its quality. It was a round pool as I saw in the half-light, entering narrowly under low-hanging trees before widening into The Wheel. The tail pulled away under more trees and the bank opposite the only fishing stand was a low cliff with trees atop. At the fishing position was more rock close behind a little beach. The current gyrated slowly round what looked like a black pit before sliding out of sight downstream. 'I'd better show you where to fish,' said my friend – 'it's a bit tricky for a stranger in the dark.' He led me down to the narrow beach and from there out among some big but fairly level stones until I was about five yards into the pool. 'Mind how you go,' he said and then with an injunction to fish out to the cliff and 'let 'em swing round' to the tail, he left me, to fish the next pool below. I was tolerably happy at first though I do not like wading unknown water at night. I had three one-pounders very quickly and then wanted to cover more water by moving nearer the tail. But my long net-handle could find no bottom in front and I was evidently right on the lip of the rock. I suddenly decided I had had enough of that and turned round rather carefully to go back; as I did so one foot found no hold and I stopped dead with both feet close together on my rock. The net shaft found no bottom to either right or left and clearly I was perched at the end of a narrow pier of rock, eighteen inches under the surface and barely that in width with deep water on three sides. I loved my friend dearly but his ears must have burned enough to stop his fishing, for he appeared as I regained the shore. 'Well,' he said when I had found some suitable words, 'I *did* tell you it was tricky.'

Far from there, down in North Devon, I was one night fishing alone in a pool somewhat similar to The Wheel. Its central point is remarkably similar in fact, a deep round hole under a high cliff on which somehow stunted trees find roothold. The tail is different however. It is long and narrow with shingle beach on the fishing side and a rock face diminishing in height as it goes towards the extreme tail, opposite. But on this gradually lowering bank there are bigger trees and their branches reach across the narrow rather fast-flowing tail to form a cavern of darkness. I know this pool intimately and exactly where the seatrout lie and take at night. Fishing off the shingle and into the blackness I hooked a seatrout which came upstream and behaved perfectly normally until suddenly it became fixed, the line went dead and I knew I was snagged. An even steady strain in a moment produced a faint wriggle; the fish was still there and there was a certain 'give' in whatever was holding it. I waded in on shifting gravel having put the rod down in a safe place, drew off some line and bit by bit shortened line from the rod top until the tip of a twig appeared. This secured I repeated the process until a thin branch emerged to be grasped in its turn. It required patience and inch-by-inch progress in working down the branch. Water was within an inch of my wader tops and I was leaning forward to the limit of balance when a six-foot length of branch came free from its anchorage and in the tangle of it was my seatrout. I was starting to back out of the water dragging the branch and its heavy load of rubbish with me when I felt a sharp blow on my shins, a salmon shot out of the water a yard upstream from me and I sat down heavily in two feet of water. The seatrout was two-and-a-half pounds and quite dead. I took it and my dripping carcase home conscious that for the first and only time in my life I had been run down by a salmon.

If you think that is simply a fisherman's yarn, it is not. It is a night fisherman's yarn and it happens to be perfectly true.

XIII

To Land a Fish

THERE is a movement, still gaining momentum, against the use of the gaff and thereby many kelts will be saved to swim away without that fatal loss of blood. My own views on the usefulness of saving spawned fish are expressed in Chapter XXVII but I may easily be quite wrong. However, if we are going to err, let us do so on the safe side. There is no excuse, really, for gaffing the average kelt though it used to be done regularly – there may be some reason in it when the fish is one of those few which are difficult to classify without a long close look.

Kelts have come into this only because their saving is the most obvious direct benefit from banishment of the big hook. There are others which include non-disfigurement of the fish and, if you like, avoidance of waste since the gaff wound almost always suffuses with blood an area in one of the best 'cuts' of the fish. By no means least is the fact that the standard of skill with the gaff is less than that with any other landing implement. To invite a passer-by to assist is usually fatal and beginners can cause disaster so easily from over-anxiety. One of the things in fishing hardest to describe is the strength of the movement necessary to sink the gaff point, but it is simple to instruct anyone to submerge a net and lift it when you give the word. I used to use a gaff but the row of different types in my equipment rack have not been used for years and are retired for my lifetime. I do know a few places in Norway where it could be difficult to take a fish from extremely strong water with anything but a long gaff and it is only in such places that I would permit my gillie to use one.

The increasing use of nets is an excellent move and I am personally happy to see more and more in use every season. This seems to follow the development of nets that are big enough to land salmon easily but light enough for the unattended fisherman to carry on the river and use one-handed there or in the boat. Some of the older types were heavy and cumbersome and if of the folding arm kind, either jammed at the knuckle because the alloy of which they were made corroded after wetting and lost its lubrication, or they nipped and so severed the retaining cord so that the net slipped down and off its arm. Nothing beats a big round frame of light metal sliding on a square section metal shaft. This net can be carried in a sling on the back for bank fishing and is ready for action by tilting head down so that the net-ring slides to the end of the shaft. The ring can be twenty-two or twenty-four inches in diameter and will handle any but very big salmon indeed. The net itself should be really deep – some I see about are too shallow. A big fish can be handled in a relatively small net so long as it is deep.

Some gillies who have an odd disregard for the care of gear have a habit of using the net as a carrier of fish between river or boat and car or house. I will not allow it because it subjects net and frame to strain they were never intended to withstand. I have seen a gillie put three ten-pound fish in the net and rest the shaft on his shoulder to carry the load a few hundred yards. All the cords of the net would be strained with the *deadweight* of the fish and the shaft could be distorted. Three fish is certainly an awkward burden for one man but in such case the fisherman should be well content to carry one of his fish. Again about nets, have them dyed green or brown, never the bright yellow or straw colour one so often sees. The most dangerous feature of a salmon's resistance is the final flurry, the last dash for freedom, and that is easily stimulated if you push a brightly-coloured net towards him.

Finally, nets rot. Two or three seasons ago I watched a fisherman playing a salmon with line through the hoop of the net and a more awkward business cannot be imagined, mainly because of the dropper fly and of course the fish had had a rest and was quite active again. All this because the net was rotten and the fish simply fell through when lifted. The subsequent inquest – a little hilarious – showed that the owner had no idea how many years that net had been on the frame. Suspicion aroused, I insisted that evening on an inspection of the rest of his gear which revealed two lines as rotten as the net. Incidentally, this illustrated how light a strain will kill a salmon on occasion: one of the lines had held a fish the previous day and I had been a spectator urging the angler to give the fish some 'stick'. Happily my advice was ignored and the play proceeded with the rod hardly flexed by my standards. The line, tested after the net affair, broke off in small pieces at the lightest pull! Not guilty of this sort of thing, I do nevertheless find myself prone to forget how long nets have been in service. Seasons pass quickly and a net which one idly regards as 'new' may be five or six years old. One must just test frequently.

Tailers are in common use in some districts and the habit of using them in preference to other devices is curiously local. The West of England seems to be one of these and on some rivers there almost every angler, outside seatrout time, carries a tailer. I have never owned one because I dislike this way of securing a fish. My mistrust is probably groundless and I have never had any trouble when asked by a friend to tail a fish. But I would much sooner do so with my bare hand and habitually do so in suitable circumstances or in the absence of the net. The method has the great merit of simplicity and of having only one restrictive rule – that which dictates that the fish shall be played out before any attempt is made. The net scores slightly here, of course, because one often enough gets a fair chance of netting before the fish is really 'done'. A similar chance of hand-tailing *can* occur but the attendant risk is far greater.

Long ago I began to record in my fishing diary the manner of landing or boating of each salmon, marking the entries 'gaffed', 'netted', 'hand-tailed' or 'beached' and on multiple fish days indicating totals so treated. It has been interesting to glance through the pages since I began to write this chapter and to see that there are as many hand-tailed or beached as secured by other means: I have neither record nor recollection of any fish lost as a result but I remember a few mishaps with the net and an odd fish scratched by the gaff.

The prime requirements for successful hand-tailing are courage and calmness once the rule I have mentioned has been observed and the fish is well-played out and quiet or moving slowly and weakly. It is easier from a sitting position in a boat than from shore or wading because you can get so much nearer to the fish as a rule, and you have no current to cause the fish to sway or swing away at the critical time. It is hard for a right-handed person to envisage the degree of dexterity of someone left-handed but I see no reason why the left hand should not be used. Right-handed, I use that hand to grasp the fish, retaining the rod in my left where it has been since the fish was hooked. When the fish is ready for tailing I shorten line enough to enable me to bring the fish close – putting the rod back and keeping a few feet of line loose but held by the pressure of a finger against the rod. If I do in some way muff the tailing the line can be released at once and the rod brought forward again to allow line to feed off the reel if the fish goes away. If all is well and the fish lies still, I reach out with my bare hand, grasp him firmly by the wrist of the tail and lift him into boat or ashore. He will not kick once his tail is clear of the water. I have read about, seen used and heard of procedures where the hand is covered with a handkerchief or dipped in sand and I have read most elaborate instructions on how the hand should be turned but I have never found a necessity for any of it. If you grasp the fish firmly – there is no need to squeeze hard, with the wrist or root of the tail closely fitting into the deep notch between fore-finger and thumb, you will have no trouble. The tail of a salmon will not fold unless it is a very small fish indeed which you should not, in any case, try to tail, any more than you would a seatrout because of the tail-folding characteristic of that fish.

With practice and the resulting confidence, you can hand-tail a salmon before he is beaten to the turning-over and lying-still stage and I do not hesitate to land swimming fish so long as they are moving *slowly*. If you can get a fish nearly enough beaten to swim slowly beside the boat or near you in the river, you can take him if you are quite unhurried. At this game you must not snatch. If the fish is sufficiently tired you can reach out, pass fingers lightly along his back and tighten as you reach the wrist. You can go directly to the latter but with a moving fish I find it aids timing to 'measure' him with fingers before tightening. If at your touch he shies away, he is not ready and he would have shied away anyhow had you gone directly to an attempt to grasp the tail.

Although, as I have said, tailing from a boat is easiest because of one's nearness to the victim, it contains a hazard peculiar to loch fishing – the hook of the dropper. If the dropper is six feet up the nine- or ten-foot cast, as it should be, it is well above one's head as the fish comes in on a short line by the boat – if the fish is hooked on the dropper the tail is trailing out of harm's way. There is, however, real risk in both cases if the first attempt fails and the fish slips from your grasp. Then you must take hand and arm away and crouch low very quickly. Unless you are really alert you may find yourself tethered to a pulling fish and a hook going even deeper into your hand. I have seen this happen to a gillie trying his hand at tailing and being unsure at the critical moment. The frightened salmon gave a weak little flurry and bored a yard or two away before giving in again – stunned, I think, by the language of a man with a number four Black Fairy in the ball of a horny thumb.

I mentioned beaching, and this is one of the safest ways of landing a fish. My own view is that it requires rather more skill – perhaps judgment is a better word – than hand-tailing. Beaching requires that a fish shall be beaten to the stage where it can just swim weakly. You can hand-tail a fish in one of its quiet moments but you cannot beach it before it is 'cooked'. Salmon, feeling the water too shallow, will lash wildly and dart out for deeper water and safety: they have to lack the strength to lash about before they can safely be grounded.

As soon as I can sense that a fish is well-hooked and, barring accidents, likely to come to hand in due time, I begin to look for a beaching place. Some pools or lake shores are obviously unsuitable but it may be practicable to lead a beaten fish a little way to some small shelving bank or beach. Pebbly or sandy or grassy shores will do and they abound on some waters – in others one looks vainly for them and it is useless to try to beach in the absence of a shelving shoreline. The practice of leading a fish in search of such a place can be overdone too, and I never go beyond a few yards. Since beaching requires a beaten fish it follows that he has been on somewhat longer than the fish destined for gaff, net or tailer and the hook hold is that much more worn.

I start the process of beaching a fish when, sufficiently tired, he is coming towards me on a shortening line and, after a rearward look to see that there are no fallen branches, high tussocks, rabbit burrows or other traps, step backwards with the rod up. The fish is led ashore but the real impetus comes from his own power and he literally swims himself on to dry land. As soon as his nose touches and he turns on his side, the head has to be given a little lift and he is drawn higher until his shoulders are beginning to emerge. If he kicks a little he will only push himself farther in, and in any case the air will anaesthetize him so that he lies still. At this stage the rod can be laid down, *reel handle uppermost* and point towards the water, while you walk forward to pick up the fish by the wrist of the tail. Take him well away from the water and apply the priest before unhooking! If, like me, you do not carry a priest, remember to pick up a suitable stone when you pick up the fish.

It can be helpful, in preparation for the final phase, to dig a groove or miniature trench in sand or gravel with a boot-heel and later to lead the fish into this, but it is not really necessary. A fish will slide much more easily on to grass or snow than onto dry stones but whatever the surface it is at the point when it emerges from the water that judgment is required to exert just the right amount of strain. In the water it requires little more than guidance from the taut line – too much of its deadweight in air will not do the rod any good at all.

Conditions vary widely and if you are dealing with a heavy fish which can only be manœuvred nose to stones or bank, or with any fish in a place where current is strong enough to wash him away, it is best to keep the rod in the left hand and pointing upstream as you go forward to the fish. Where you cannot for some reason easily get the head out of water it is vital that the fish should turn over and lie flat as or before his nose touches, and so slide in to contact on his side.

If my hands are very hot I usually dip my right hand in the water before grasping the fish which may kick at the touch of a relatively hot hand but I seldom do so in a boat on the lough where one's hands are rarely so warm as in bank fishing. This is a small point and I do not think there is much in it – the action of grasping the fish is a matter of instant contact and if you do not have him at once – hot hand or not – you have to let him sail round and try again.

Salmon, once beached, can be lifted by the gills, that is to say with a finger or fingers under the gill covers. This is perfectly safe so long as the fingers are really well in – preferably under the lower arc of the cover under the jaw where cover joins throat. For this the fish has to be utterly done, lying on its side and quite quiet because lifting a gill cover and getting hold this way takes longer than tailing and any fish other than nearly dead, will kick when the gill cover is lifted. And almost all – however far gone – will kick to some extent when lifted this way. I have also before now grasped a smallish fish across the top of the head with thumb under one gill and forefinger beneath the other. An iron grasp is needed and this method is not one to be recommended because the hold can be insecure.

In writing of tailing and beaching I have been dealing with fish taken on the fly but, naturally, the method can be used when spinning. There is, however, in securing a fish hooked on spinning gear one risk which does not exist with the more lissom fly rod. I use, when spinning in coloured water – I do not spin in other conditions – an eight-foot built cane bait caster with a multiplying reel and this outfit is stiffly actioned. I am always rather gentle with a fish in the end stages of a fight because so often a final flurry with this relatively unyielding rod can result in pulled out hooks – or even a broken trace. Thus if you are beaching, or preparing to land a fish on a short spinning rod, make sure he is really played to a standstill so that there is no risk of a jerking, lashing flurry as he comes in. It is worth reflecting too that a fish with a mouthful of spinning bait and triangles is not the fish you should lift by the gills!

I have written at length about some ways and means of securing salmon and about some details of them which are doubtless well-known, at risk of boring the well-equipped and the proficient. I can only say that moving among anglers in fishing quarters up and down the country, I see more inadequacy in fish-landing tackle than in any other section of equipment and less versatility in its use or non-use than in casting a fly or a bait.

XIV

The Tree

IT HAD been raining for twenty-four hours, curtains of rain driving in from a grey south-western sea across the sodden Devon countryside. Rivers ran red or grey or brown according to the soil colour of their gathering grounds. Every small grassy runnel, every ditch was making its contribution and from my study window I could see shining threads everywhere in the fields below. The tiny brook beyond the hedge had grown to the dignity of a little river and was making its voice heard. I wished that I could always hear its chuckling – the sound of a river in the night is one of the great things in nature.

It was March, but the rain was warm and in spite of the cold ground on the high moors the spate would be warm and I knew a river where fish ran early. In this river, besides, there was always the chance, if there had been plenty of water, of a February runner. Some Devon rivers, granted the water, get a scattering of such fish – mostly large spring fish of eighteen to twenty-five pounds. I can recall one year when, beginning to fish as soon as the season opened on March first, I had five fish in the first three weeks, all of them within ounces of nineteen pounds. Fish of this class are hard to find because they seldom show and there are not many of them. You have to fish the water thoroughly:

the fish enter and settle in high-water conditions and for this reason I seem to have found a high proportion of those I have had in unusual places. One I remember came from dead water under overhanging bushes and another from what is normally a kelt lie very low down in a pool tail. It must be admitted that they do not come well to a fly – even to a big one fished deep and very slowly. But, since water is *mostly* high and coloured in March, one spins with a Devon of suitable size and colour and with an easy conscience. A March drought, which can happen even in the wet West Country, is a thing to be dreaded because the small spring fish will not have been able to run and one is driven to fish for the few big early fish in low, crystal clear water and with appropriate casts and flies. The banks are dusted with celandines and there are primroses, catkins on the hazels and if the afternoon is warm a short hatch that will bring up the trout – but usually that is all. This low clear river is wrong for the season and it is a lucky man indeed who has a fish at the end of the day.

So one rejoices in a March spate in the prospect of fish fresh from the sea and for the chance it gives to show a bait in the right bait conditions, to one of the big early runners. On the day I have in mind, the rain eased with a shift to west in the wind in the late afternoon and at dusk the ditches were emptying. Main water courses would hold their water height for a few hours but by morning would begin to fall to a level they would then hold for several days – water each day clearer than the day before. I put the rods and the rest of the gear in the car.

'High and dirty,' was the report in answer to a telephone inquiry next morning, . . . 'leave it till tomorrow.'

I did. Fish will not run anyhow at the peak of any biggish flood. To undertake a long drive to find utterly hopeless conditions is a nonsense and, although there have been some notable exceptions in my fishing, I find that the selected days overall produce the better results.

When I did arrive at the river on my chosen day I found the water so heavy that bottom contours were totally obscured and shallows lost. Pools were fringed with submerged bushes and every corner carried its rotating mat of drift and rubbish. But it was falling and clearing and a sandy-grey in colour. I put the net down and found it disappeared entirely from sight eight inches or so under the surface. That, with the weight of water where one would wish to hang a fly, decided me against the fly and I took the fly gear back to the car returning with an eight-foot cane spinner, the Pflueger multiplier and a box of baits. A two-inch blue and silver Devon went on the ten-pound wire trace, itself thirty inches long with a Hillman lead on the line end of the top swivel. I am very fond of this outfit, the light single-handed rod and multiplier putting any compact, streamlined bait out like a bullet and with great accuracy. This accuracy depends only in part on casting technique – the rest on reel adjustment and on trace length. I seem to see devoted fly fishers, spinning like myself only occasionally, having difficulties directly related to the use of traces too long – but this is not a course in bait casting.

At normal height the first two hundred yards of my chosen beat was rather thin water; today it brimmed with three or four feet of boiling current and probable resting places were few. I fished the quieter margins where trifling bank irregularities made possible holds on the edge of trailing grasses. Upstream, beyond the shallow was a lovely smooth pool tail where I have had many seatrout in summer dusks, and beyond that again, the long still body of a huge pool. Still it certainly is when water is lower, now it coiled and twisted and swirled down its entire length and it must hold a fish or two. I doubted whether a fish could maintain its position in the tail, so heavy was the sweep of current, but I fished it finding that the bait worked perfectly with aid from line recovery. It was child's play. Cast straight across, recover a little line to keep the bait off the bottom in the first few yards of its progress and then leave the reel alone while the bait swung in its arc down and across to finish under my bank. Hang it there for a moment before recovering line – nothing. I had covered the tail when I met an angler acquaintance coming down my bank. It was two-thirty. He had fished four hours without moving anything. 'But,' said he, 'there is one fish in the water if you have a mind to risk your gear in trying for him. He's under the branches of that fallen tree – I've seen him twice.'

Upstream, where he pointed, a great ash had fallen across the river from the far bank. I hitched the bait up and went to look more closely. The tree lay a third of the way down the pool and at its deepest part. The mossy trunk was half-submerged and the topmost branches well out into the fields on my side, while tons of earth and roots adorned the butt. I could picture the River Board clearance gang scratching their heads over this one. The current surged among the branches and boiled up under the trunk on the downstream side and here, as above, was a forest of branches. It was here, among them, that a fish had been showing. A worse looking place I had never seen. The trouble was that I had no means of knowing how far the branches extended – which were broken off and what other debris was tangled up down there. It was too much to expect that a bait put across within fifteen or twenty feet of the trunk would come through yet it was here that the fish had shown twice. I spread my game bag on the sodden grass and sat there for a few minutes. The cold and blustering west wind was falling light for the time and was no longer pushing dark catspaws across the swirling river; it was perceptibly warmer. I knew that if there was to be any taking-time on this day it would be during the softer spell.

The first retrieve brought the bait back to my feet smoothly enough but the second ended in a vibrant jar half-way across the river and I could visualize the triangle over the barbs in one of the hidden branches. But the 'branch' at once shifted its position down and away and I had my fish – a very strong one. I stayed above him fearful lest a downstream pull should urge him up and into the forest above. A peril almost as great lay downstream although further away because if he had gone down to the tail I could never have held him in that racing glide of heavy water. So I held him firmly but not hard, twenty yards below the tree and after rather a long interval of circling and boring and short diving runs, he came in. He was bigger than I had thought and now I had the problem of getting him out. The bank was very steep and fringed with outward-growing bushes which, in a low river, would back a little stony beach. There were, however, small gaps, one just below me. The fish was on the surface swaying wearily and almost incapable of pulling hard again but it was a nasty place because if he did turn aside, or was swept downstream while close in, he would be under the bushes and I would be lucky to get him back. It had to be done quickly but without flurry. When he was ready for the net he lay finning the current, head upstream. I pulled him a little closer and he turned head to bank and put his nose under a crossing root in the water at my feet! Five or six inches of his head were now under the branch and no outward strain I could apply would force him out because I no longer had a direct pull on him; I could see the wire of the trace cutting into the root. There was only one thing for it. I had to stretch to my limit to get the big net under him tail first – a move I dislike but the current streamed out the net – and trust that his head and shoulders would drop back as I started to lift. It was, in the end, a combination of outward pull from the rod-tip and lift of the net rim that got all of him into the bag of the net. I dropped the rod across the bushes; the fish felt like twenty pounds, looked it laid on the grass, and was in fact precisely that on the scales that evening. Fresh and bright he was as beautiful as the first fish of the season ought to be. I did a little brow-mopping, smoked a cigarette, looked at the fish and the river and reflected that it was good to be alive.

It was three-thirty and the mild air persisted. I decided to fish the rest of the pool before calling it a day and replaced the wire trace which was rubbed bright in places and could have been kinked. The triangle on the bait's tail was kinked to one side as it so often is after playing a fish and I struggled to get it straight again – finally discarding it altogether. I like a bait for salmon to spin slowly but quite straightly without wobble.

Luck is not indefinitely elastic and I chose this time to stay well below the tree and any possible snags associated with it. Thus the bait would fish the deepest part of the pool where there was a known lie holding some fish at normal river level. About the fourth or fifth time over the bait stopped dead in mid-stream three or four yards below where I calculated the tips of the sunken branches would be and I was into another fish. This one was a head-shaker and I thought he was probably hooked either in the tongue or was lipped so that he could see or feel the bait outside his mouth. Anyhow he went jagging across the pool giving me slow convulsive pulls as if he were a very heavy fish. Again with one eye on the tremendous current pull in the tail I stayed opposite him and managed to keep in the deep centre. He made two quite irresistible runs to the far bank, threshing along close to the smooth clay and showing bright gleams of his broad sides as he rolled over and went deep again. One run left me with a burnt thumb from rather excessive braking on the reel drum, but in the end he came sailing quietly and slowly over the net and I had him. On the grass beside the first fish, he was its twin; both were cocks – the first perhaps two or three weeks in the river, the second obviously in on the present flood. I labelled them both twenty pounds but later found the fresher fish to be nineteen-and-a-half.

My fair share of fish for that day lay on the bank, the wind had come up again, little dark rain clouds were flocculating in the west, and the brief warm spell was over. It was late afternoon of a March day again. It is strange – and yet *not* strange when you know the ways of salmon – that the chance of a fish comes with a change of some kind, a softer air, a different light. It was especially interesting on this day because in my experience, the afternoon is seldom as good as the morning when one is spinning. I have never known why this should be so but

I would rather have one hour before two o'clock (sun time) than two afterwards, for spinning.

A dismantled rod, even a two-piece, without its case is a nuisance. If you hold it firmly the sections rub one against the other to damage the finish and if you hold loosely one piece goes askew and can stab the ground or become entangled with one of the traps that bedevil a laden angler. I slung bag and net, tucked the rod under one armpit and hooked fingers of each hand into the gills of a fish. And so to the distant car, with rests to ease slimy cramped fingers and stretched arms. The day was done.

One thing more. The first fish, with his nose under the crossing root was the second salmon I have seen perform this trick. On both occasions the fish stopped of his own volition with about four inches of his nose projecting beyond the root and there stayed. It makes me wonder how often fish which are apparently immovable somewhere out of sight of the fisherman are doing just this.

XV

Tackle for Today

IT HAS occurred to me that whilst rods and tackle have been brought into accord with modern fishing ideas and practice – written and restrained guidance on equipment has not. There are some excellent books on the last big change in salmon fishing technique – the introduction of the greased line, but very little indeed finds its way into print on the gradual change in conventional sunk line gear which has been proceeding for many years and is still under way in the sense that more anglers conform to it every season. The change is towards lightness.

If you fish, or judge current ideas by what you see on every big river, you may doubt the degree of change in rods because there a physical limitation is at work to prevent moves to lighter stuff. Whatever your capacity to throw and fish a long line, there are rivers and pools on them where the short rod in common use today simply will not either cover the water or fish the fly and, of course, there is no point in pretending that a really heavy fish can be held in a big river's strong water with anything less than twelve feet of length and/or the backbone that goes with normal taper of cane in that length. There are after all, times and places where you have to hold a fish very hard indeed and to the limits of

PLATE 6 QUIET WEATHER MEANS IDLE BOATS AND IMPATIENT ANGLERS

PLATE 7

A TEN POUNDER COMES TO THE NET
—PLAYED OUT

the tackle. For this reason then, you may still see the big rod up to perhaps fourteen feet in use where it fits the need and where the lighter rod, which I contend suits the bulk of salmon fishing, would be laughed at even before it proved its insufficiency.

From the beginnings of salmon fishing for sport down to today, rods have changed in three fairly well-defined stages. First we had the enormous timber 'poles' of eighteen and twenty feet which have always puzzled me because on the basis of handling two of them and fishing with one, I could never understand what kind of men these early fishers were. A day with one of these rods would be quite beyond the powers of most of us and I cannot imagine wherein lay the pleasure. Next, with a change to greenheart, lancewood or split cane came the big salmon rod of my own boyhood – frequently sixteen feet and sometimes steel-centred. At fourteen feet they were heavier than a modern rod of similar length but eminently fishable; the sixteen-feet specimens were mostly brutes and I have ended many youthful days with one of these very tired indeed. Master craftsmen did, however, produce some greenheart rods of this generation which had a beautiful even action and were a joy to fish once you knew how to let them do rather more of the work than the average rod would then or will now do. It would be ungracious not to bow to the skill of the men, themselves fishermen, who made them.

So-called light rods of cane, in the twelve- to thirteen-feet category, came in with greased-line fishing and this 'fashion' represents the most recent marked change in rods. The shift to *truly* light rods began somewhere about that time, it seems to me, and as I have said has spread gradually and is still doing so. It may be that the greased-line method, using what in the previous generation was known as a 'grilse' rod, made fishermen think by perfectly logical steps from light two-handed rods to the single-handed rod in wide use for general salmon fishing in recent years.

I have often been asked what kind of outfit I would suggest for an angler taking up salmon fishing again after a lapse of years abroad or of non-fishing for some other reason, or for a beginner. It is important for the first man to realize the possible extent of change in ideas since he last fished and for the second to know that he will double his pleasure in the fishing if he avoids the big rod. The review of changing conceptions about rods is I think a proper and fitting lead to any advice about going in for lightness and I have dilated about early rods for that reason. The choice of rod and running tackle should be governed in the main by the kind of fishing which is planned or anticipated and by the consideration that lighter lines and casts and smaller flies are in use now than were used even ten years ago. Line change springs from rod change and cast and fly modification from the difficulty of taking salmon in the manner of former years. Chapter III includes some thoughts on this rather perplexing question.

It is worth reflecting that scores of salmon are taken every season on trout tackle, mostly at the end of rather long fights which cannot be pressed to shortening without risk of tackle breakage or damage to rods too slim in build for the job. The fact no more than points the way to the sensible medium and this is what I would have my inquirer buy for his salmon fishing in small to medium rivers and for all lake fishing. I have taken many salmon while seatrout fishing on little trout rods of nine feet and under and it has almost always been a long job, first because the little rod simply was not 'man enough' and second because the reel to balance did not carry enough line. This latter meant, naturally, that I could not press the fish hard enough for fear of provoking rough and unmanageable responses in the form of long runs.

For these reasons despite transatlantic fashion in rods of few feet and fewer ounces – my sensible medium is a rather powerful ten-footer or best of all, in my considered opinion, an eleven-footer, cane of course. I will have none of the smaller and lighter style for salmon. A rod of eleven feet is my own standard equipment today as it has been for some years. It has handled some hundreds of fish but I will admit that I seldom fish big rivers or big fish rivers. My fish average ten pounds and I find that the eleven-foot rod handles the occasional twenty-pounder very well – as well as my now discarded twelves and thirteens – while giving me some sport from the odd little fish of five or six pounds. It fishes single-handed, is long enough to dibble the dropper in lough fishing, and will carry a reel of adequate line capacity.

In three pieces, it weights ten ounces and is balanced by a three-and-a-quarter-inch reel weighing, with line and backing, twelve ounces. I do not specify the size designation of the line, which can be of the sinking variety and greased when you want to fish greased line, or of the modern self-buoyant type. I say only that the line should bring out the action of the rod, and that most buyers choose lines *too light*. I use a double-tapered silk line of the solid, that is sinking, kind. The taper is not necessary but I like to see the line go down nicely and the taper helps. Rods vary widely and the only way to choose reel and line to suit any rod is to try several combinations until you find the one that feels right to your wrist and manner of casting. I insist only that you should have forty yards of running line and one hundred yards of backing. So-equipped you can fish confidently in nine out of ten rivers and all lochs.

The eleven-foot rod so often seen some fifty years ago, in Scotland and Northern Ireland, has never for some reason been popular in England and I have not seen it in Wales but it has always been a supremely *useful* rod. Some of the early greenhearts – I had several in my growing years – were delightful to fish with and at eleven feet one of these gave you more versatility and as much command of water, on occasion, as bigger and less handy weapons. This is equally true of one in good split-cane today and is why it is my own favourite and my recommendation for you if you are starting to fish salmon again – or for the first time.

I hope that I have not made an unwarranted assumption in dealing at length with fly gear only – and I hope that choice of spinning gear is of secondary interest. I have nothing against spinning as a part of a fisherman's craft to be used on the *rare* occasions when fly is unsuitable, in heavy or dirty floods or in early spring when even a very big fly will not go deep enough. I wrote at length about the evils of indiscriminate spinning in an earlier book* and I will only say here that fishermen of my generation – the 'squares' of angling – cannot help being saddened by the launching of so many new anglers on a career of spinning, at all times, places and conditions of water and weather. An appalling amount of damage is so done.

* *Salmon and Seatrout in Wild Places*

Agreeing, however, that a spinning outfit is a proper component in the equipment of a salmon fisher for use in suitable conditions, I would suggest that you avoid the ultra-modern in fixed spool reels and rods to match. They are almost essential for very light spinning, which is precisely the method which can be so damaging to stocks of immature fish. For medium and heavy spinning I think that cane rods of medium build and nine or ten feet in length for use double-handed, or of eight feet single-handed are right in conjunction with the multiplying reel – preferably the 'Supreme' which has for thirty years or more been in my view the world's finest spinning reel. Braided terylene line can now be had in very small diameters and a hundred-and-fifty yards of fifteen pounds' breaking-strain will go on one of these reels. The traces used with this line should break before the line parts in the event of snagging and ten-pound nylon or wire is suitable. The nylon monofilament line used on fixed-spool reels will not cast well off a multiplier and if you think this a rather unnecessary comment I must tell you that I know of two cases in which beginners had been started off with this type of line on multipliers. Most dealers will not, of course, do this.

Back to fly work again and on the subject of light rods and proportionate end tackle it seems there is little that can usefully be said about casts. They must always be selected to suit conditions but the very sensible move towards the use of lighter and often smaller flies which has grown up over the years naturally dictates cast thickness. Little flies on stout casts are as ridiculous as big ones on light casts. The flies have to 'work' and they will only do so when cast and flies are in balance, they also have to *feel* right in both back and forward casts and

I think this in an odd way is the acid test when optimum line length is out.

River fishing permits, and on occasion demands, the use of a greater range in fly size than loch fishing but for all fishing I believe that most men fish too big for most conditions. Size is especially critical in lochs and unbroken water and if you are going to fish this type of water to any extent it is an expensive mistake to fill boxes with flies bigger than fours and the bulk should be sevens or eights. A few bigger flies for heavy water are in order but their usefulness is limited. Any modern range should, however, include some of the hair-winged variety whether on conventional hooks or dressed as tubes. The small sharp triangles fitted to the latter are very good hookers and holders. About long-shanked, fine wire, low water dressings I am not absolutely sure. I once saw an epidemic of so-called 'short-rising' and non-hooking apparently cured by the use of long-shanked hooks but have never been quite sure whether the hooks *were* more effective, or whether the mood of the fish changed. Short risers among salmon do not come as close to the fly as most people think: the visible evidence is caused by the fish turning away and he may easily not have come within a foot of the fly. Many fish doing this are merely attracted by the moving fly: they only come to look at it and their mouths are closed. The type of hook makes no difference – what is needed is another size of dressing and/or another pattern.

The tendency towards the use of smaller flies is not a figment of my imagination though it is not growing as fast as I think it should and by no means in line with the results of those who think of a size and then fish one size smaller. So very much, as in all the elements of fishing, depends upon conditions and he would be both foolish and unkind who would be too dogmatic in endeavouring to guide choice of equipment.

For my part I have made some suggestions based on experience so long that I can now count my remaining years of fishing. If you find them unacceptable do, I beg you, at least err if err you must, on the side of lightness of rod, of fine terminal gear and of small flies. They all suit salmon fishing today.

XVI

On Strange Water

HAVING rented, begged or borrowed some fishing which is new to me, and driven a very long way to get to it how, on arrival, do I start – where and with what flies do I begin? This question was put to me quite seriously in something like that form, and of course it is impossible to answer, at least in concise but comprehensive terms. Some assumptions have to be made, first because of the reference to long driving that the fishing is in Scotland, Ireland or remote Wales and second that it is seatrout fishing. Trout fishing would have to be magnificent to justify the journey though the idea is tenable, and I can think of few salmon waters open to the visitor where basic information would be lacking from the outset, but it *is* conceivable that the enterprise could concern seatrout. In process of compressing the question to make it capable of answer I have to make the further assumption that loch, lough or lake fishing is involved, though that eliminates Wales.

Fishermen in general are much more knowledgeable about river fishing and a strange river is in so many ways a far easier proposition to tackle than a strange lake. Seatrout are, of course, queer, moody, fickle fish in their taking but their preference for type of water – pool or run, glide and stickle, sheltered and open water – hardly varies from river to river. Very small streams present special problems in approach but, all in all, I think it is big sheets of water that may tend to appal the stranger. And I have known it happen that accommodation and fishing have been booked on the strength of a guide-book or publicity comment like 'Seatrout run from the middle of July'. I do not think I would go anywhere, except possibly to the West of Ireland, on the basis of such scanty data but I do meet anglers in Scotland and Ireland – usually the only angling members of a family holiday party – who have done so. Their difficulties are not eased by their status as fishers on odd days because the best sources of local information are the pick of the available gillies and these are invariably snapped up by the dedicated ones who have come to fish – and only to fish – every day and all of every day. There may be an opportunity to share a boat with a man who either knows the water well or has a good gillie but so often if happens that the gardener or stableman, probably willing but the reverse of knowledgeable, is pressed into service to boat for the stranger.

Not so long ago in Ireland I was coming in from a day with the seatrout on a big lough, trudging up from the boat and looking forward to my drink, my bath and change and my dinner, when I met a new arrival heading for the boat-landing with the household handyman in attendance as gillie. He had had an early dinner and was out for the 'evening rise'. We had some seatrout to show for our day but none during the last hour before coming in and neither my gillie nor me had the heart to tell the newcomer that on this lough there is no such thing as an evening rise. I did tell him later, when returned puzzled and disappointed, for it was a warm July evening with a nice breeze and in other loughs that I could list, fish *would* be taking. This sort of thing which can be related not only to fishing time as in this case, but to choice of fishing grounds, of flies and style of fishing, can so erode results that days are wasted. Any first visit must involve some apparent waste of fishing time: I consider, after long years of exploration, that it takes about six weeks of fishing to begin to understand a lough – and more if it is a really big one with scattered fishing grounds.

Not everyone has either patience or leisure for such a programme and a fair and accurate, if somewhat heartless, answer to my questioner could have been: 'If you cannot secure in advance the services of a good gillie or otherwise, again in advance, make contact with some reliable source of information, don't go.' This should be an inflexible rule when the object is fishing and only fishing. It is still sound but less vital when fishing is only one proposed activity out of several.

However, there are some things which can be said to assist on the odd occasion when, by force of circumstances, information is nil and one has to operate from scratch – faced with a strange loch but sustained by the knowledge that seatrout are 'in'. This last *has* to be established by local inquiry on arrival in the area because it is the one point on which advance information may be of limited value. Your contact may tell you, a week before you are due – 'no seatrout yet but they're in the estuary and should come in on the next spate. After three dry weeks it looks like rain this evening . . .' You go, high in hope, and you do not get the rain and the seatrout trickle through in penny numbers: this has happened to me quite often. The only advance intelligence that is really useful has to be positive – 'the fish are here'. So, the actual presence of seatrout in numbers really has to be confirmed on the spot.

We come to the point where you have arrived and where it is known that the fish are there – but that is all. You do not know precisely where the fish lie and can be made to rise, nor what flies to use. Equipment is simple and not likely to present the average angler – even the occasional one – with any difficulty. A nine- or ten-foot trout rod will serve but I would like more line on the reel than is needed for, say, the usual herring-sized trout. Your big new lake may hold big seatrout and, though these very 'stiff' fish are always hard to come by, some surprising things happen, and the casual visitor is as likely to get into one of them as an old local hand. If there is a choice of reels, therefore, use that which carries not less than sixty yards of line: it is not, however, desperately important and for ninety-days in a hundred a trout reel and line will do. The cast should carry two droppers or, more accurately, a dropper and a bob so that you fish three flies; this gives you three chances of finding the fly of the day or the type for the lough. Cast caliper, weight or breaking strain is not *quite* so simple to settle on because much depends on the quality of ripple you have on your fishing day and the class of fish then present. Taking the latter first, it can be said that if the finnock or whiting run is at its peak you will hook more of these than of bigger fish and the cast can be 2x or 3x or nylon of about two-and-a-half pounds to three-and-a-half pounds breaking strain. I should choose 2x. But if bigger fish are present in numbers and there is a good chance of three- or four-pounders and over, you really need 1x or five-pound nylon. In a very light ripple you will have to risk the lighter cast. Seatrout are hard pullers but with rod and reel perfectly handled they will not break 2x by this means. It is their activity and the jumping on a long sunken line that tests light tackle. I must confess that in a real breeze and wave or when big seatrout may be met, I like 1x and the cast should be three yards long.

Now to flies. Wherever your lake may be, one fly must always be on the cast and preferably in the top position, on the bob – the Butcher. The other two can be Peter Ross on the dropper and Connemara Black (which will take fish anywhere – not only in its native country) on the tail. An alternative might be Zulu on the bob – the blue variety in some parts of Scotland – Blue and Silver in the middle and Golden Olive (Ireland and some parts of Scotland) or Woodcock and Yellow or Mallard and Claret, on the tail. I believe that Mallard and Claret is more generally successful than any other fly, but that the Butcher, because it is fished more where seatrout are in greater numbers, in the north and west of these islands, takes a higher total. I was not asked for a dissertation on seatrout flies but for some 'starting' advice and I am quite certain that among those listed above the taking fly will quickly be found anywhere. My own first trial cast would always be Butcher top, Peter Ross middle and Connemara Black tail. A local pattern or patterns, which exist by the score for seatrout as they do for salmon and seatrout, may kill more fish but you cannot know of them.

Equipped, you face the second stiff problem – where to fish. To solve this in a way which will afford some, if not the best, of the sport to be had from your strange loch, you need two things which as a fisherman you probably have – a knowledge of seatrout ways and an eye for country. Country in this context means underwater country – the probable contours of the loch floor. Advance study of a big lough fishing new to me always sends me in search of Bathymetrical Survey notes which sketch a general picture of the hollow, depression, trough or valley which holds the water and gives some indication of its geological origin and subsequent modification by glacial action or erosion. With these notes, if you can locate them, and a contour map of the area you can form a very sound impression of your lough and its physical features. Superimpose your understanding of the habits of seatrout and you know where it is reasonable to look for fish.

You can make mistakes, of course, and an area which by all the rules should hold fish may, for some reason peculiar to the place, never hold any. Seatrout are not very collaborative in keeping to anglers' rules but in a day or two's trying in fair conditions I would undertake to find fish.

Seatrout are great travellers and they scatter widely. But they like certain types of ground and the deltas of inflowing streams, especially if they are clean sand or gravel, are certain finds. Often they will hang about the mouths of quite small entering burns even where the shore is high and rocky and there is no delta laid down offshore. Clean stony shores are almost always better than soft-bottomed areas but seatrout have a great liking for the proximity of reed beds. Here, naturally, the bottom is liable to be soft and may indeed be foul with rotted reeds. The fish like the nearby shelter of the reeds as is quickly proved if you hook a strong fish. In some Hebridean lochs the reeds are an ever-present menace; fish lie close to them and dive straight in on being hooked. Salmon will do this through sheer strength and blundering – seatrout do it out of innate cunning. A seatrout in a river pool knows every bit of refuge within twenty-four hours of entry – and he displays the same awareness in a loch which is, after all, only a big pool.

Island shores, unless water is very deep, are worth fishing always as are rocky points or gravelly spits, loch-head shallows of sand or gravel and in fact any clean-bottomed areas of suitable depth. In general seatrout may be induced to rise in deeper water than is ideal for salmon but it is hardly sensible to try to specify depths. I know some unusually clear lakes where good seatrout grounds carry eighteen or twenty feet of water and others, heavily peat-stained or very dark-bottomed where you ought to go inshore when you find eight feet under the keel. Again, seatrout can be risen to the fly in *very* deep water when they are really 'up' – either feeding or for some reason, perhaps temperature of water in the shallows, moving in the upper layers of deep water. Over-fishing will move them out of shallow water too. Thus, on the question of depth, you must make some assessment of the type of loch and choose ground accordingly.

There is a factor which has been left wholly out of account in these notes on choice of fishing ground and that is wind and weather. In Chapter XVIII I have dealt with some of the hazards which *can* but need not harass the stranger who is boat fishing without local help. Commonsense will dictate that some points on a loch shore which by map reference look promising, are beyond reach in some weather conditions. At such times it may be useful to have other activities – you could for example and with profit to your store of information, climb the highest nearby hill and look down on the loch. In the right light one can sometimes see the bottom and the colours of varying depths.

Salmon may be in your seatrout loch and lying in the same areas. This is a contingency which plenty of line and gentle hands can often take care of so that your salmon may be reduced into possession and a bonus on your seatrout fishing.

XVII

Lough of the Big Wind

THE glen runs east and west. It draws wind prevailingly from the Atlantic, compresses it between its two-thousand-feet-high walls and releases it where they recede, to blast half the County of Donegal. Just so long as there is some west in a gentle stir of air along the mountainsides, there will be a steady fair breeze in the glen – a strong wind over that geologically tortured landscape will transform a haven of summer mountain grandeur into a shrunken world of screaming gusts.

The floor of the glen is occupied by a narrow winding lough – a considerable area of water which because of the steepness of the hillsides has few shallow shores along its length but a big beach and fan of sand where the river enters at the western end. The water, coming off rock rather than peat, has very little colour and out from the lough head beach one can see the clean sand bottom for hundreds of yards in any direction. This forms the main, almost the only, fishing area and as the season advances carries a big stock of seatrout and some salmon. The fish gather here because until they move into the river to spawn in autumn floods, they cannot get further. One branch of the little head river comes down a high cliff face not far from the

PLATE 8

A STRONG WEST WIND, BLOWING FROM THE NARROW LOUGHHEAD IN THE LEFT BACKGROUND CAN MAKE THIS WATER TOTALLY UNFISHABLE IN MINUTES, SO THAT A SCENE OF SUMMER MOUNTAIN GRANDEUR BECOMES A SHRUNKEN WORLD OF SCREAMING STORM.

SEE CHAPTER XVII

PLATE 9

THE LOW WALL, DIAGONALLY ASCENDING THE WEIR ON THE RIGHT IS A SIMPLE BUT EFFECTIVE PASS

lough and no fish can go higher in this arm of the river; the other quickly peters out. Fish come from the sea up a boisterous river famous for its salmon, through a big lough where many again are caught, and finally through a second river, sluggishly meandering through boggy moorland to enter the lough of the big wind. Here in rarely disturbed and carefully guarded water, there can be superb fishing. Days there are favours to a very few, but the real problem is the weather – the plan to fish has to be the plan of the day only, finalized when the weather of the day can reasonably be judged. Even then to judge from a distance is to gamble heavily; one can start out in a near calm and find a rising gale on arrival and on one occasion at least this happened to me. You shall judge whether my gamble was lost or won.

This was the way of it. I had been mixing seatrout and brown trout fishing and getting pleasant little bags of browns and twos and threes of seatrout in a rough unseasonable August. The salmon in the lake were stale and dark and the lily-pads were thick along the edges of the reeds. Salmon and lily-pads do not belong together and the spring fishing ground had an unreal look despite the plunging fish. About the middle of the month I had a great urge to get among the seatrout, in a place where I could open my shoulders and cast a line. I love little rivers but I was weary for the moment of rock-climbing on their banks and of creeping through tangled gorse and wet fern. I wanted to sit out on a big lough and fish. Dedicated lough fishers will know what I mean – the heave of the boat, slow talk with the gillie and a tripping bob-fly to watch.

We picked a day when, down in the village, the

slender tips of the ash trees barely moved and the blue turf smoke just drifted, and by mid-morning we were tugging the boat foot by foot down some yards of very sticky yellow sand. The lough of the big wind was steel-grey under grey clouds but there was no more than a nice seatrout ripple over the fishing ground when we pushed off. Wind was offshore and directly down the lough. I had my eleven-foot cane and a 1x cast with two flies which I think were Silver Doctor and Golden Olive. I am not sure, but it matters not at all because they did not bring a single rise though John worked the boat cunningly in short diagonals across the shallows. A remarkably bold stag stared at us from the shore and I thought how he fitted his background – the heather at the back of the beach, the ancient wind-contorted pines in twos and threes behind that, and the mass of other trees dark against the pale rock of the mountain.

John came back over the water we had fished and started a fresh series and now with four feet of water under us I began to look for fish: the best I could find was a single pounder. We had now been once over about a quarter of the fishing area and it was clear that something was amiss. Something was, as the next few minutes showed because the weather broke as suddenly as it can among the high hills. The rain came first – biting cold rain that sent us quickly into tightly-buttoned waterproofs – and hard on its heels the wind. In minutes we had a salmon wave in place of the light ripple and the blast through the tossing pines seemed to be growing. 'John,' I said, 'we shall be lucky if we are not blown off before we've fairly started.' And my gillie, who holds this place in something like awe, good rough-water man as he is, put the boat in while we found a long, narrow, heavy stone which, with the very long mooring line knotted round its middle, went in the bow. I left the tail fly where it was but put a rather big Zulu on the bob. It was the type of Zulu I like with really wide silver tinsel prominent along the body. My diary is lacking in that it fails to record the size but I think it was a seven – which makes it a big Zulu. I thought that once waterlogged it would hold the cast down in the big blow that by all the signs was coming.

The next working line across the narrow head of the lough found us about forty yards off the beach and I would say that wind and wave were enough but not too big. I rose fish after fish mainly in the pound to pound-and-a-half class and we netted three of them. The Zulu was too big for this water and the class of seatrout we were finding, but the best drifts were yet to be fished farther out, and there the wave was bigger and the bulky fly might come into its own. Still the wind grew, and worse, was becoming squally. We put the boat head to wind and John held her in very short drifts while I fished off the stern. The stone in the bow kept her stable but every now and then would come a blast that moved us bodily, flattening and blackening the water before chasing away down the lough to whip the tops off the waves as it gathered momentum. In one of the gusts I found myself playing a rather better fish – two-and-a-half pounds – which I had to net because John dare not let the boat go. Conditions were becoming downright unpleasant: in the squalls the bitterly cold driven rain stung our hands and everything in the boat streamed with water. The line shot sluggishly as very wet lines do and the cast felt soft. There were fewer rises but those seatrout that did come up meant to take and the Zulu accounted for two more of about two pounds each.

That gave us seven and I looked at them slithering to and fro in two inches of rain water and decided it would be kind to John to call it a day. I twisted round and made appropriate gestures to him but he shook his head and motioned towards the bow with a jerk of his chin. We were no more than fifty yards off-shore but it took five minutes of hard pulling before the boat grounded. Hands partly dried on ends of scarves, we somehow got cigarettes going but they were quickly sodden and cast away. John untied the stone in the bow and measured the line: there were about seventy yards of it and it was almost new.

To our left and opposite the start of the unfished water was a solitary pine rooted in the edge of the heather and a bare two yards from the water's edge, and to this, low down, John secured the free end of the line. As we settled in the boat again we were struck by the most appalling squall. It roared through the pines behind us and out over the lough, picking off the tops of the waves and, while we watched, lifting solid sheets of water high over the shoreline trees. Two-thirds of the glen were obscured by low cloud, driving rain and water torn off the lough. No boat could have lived in it.

I thought we might become airborne at the end of our rope and I was sure of it if we ever swung broadside on. John said, 'Holy Biddy!' as we sailed out, dug his oars in hard and put his back into retarding our slide till our mooring came taut. When it did we were reasonably steady but swinging with a most disconcerting motion in a wide arc. I lifted the rod and fed some line to stream out in the gale, lowered it to parallel the water and worked the fly by simple movement to the right. The fly did not work – it merely travelled along waves and hollows – mostly rather deep. It suddenly stopped dead and I was into something – either a big seatrout or a salmon. I had not seen any break in the wave rolling away from us, nor the fish, and after two or three minutes decided we had a big seatrout. There was no salmon play in this fish – he was down on the bottom, surging about and pulling hard but not going far away. I dare not lean too hard on him because the boat swung like a pendulum when we suffered an extra buffet from the wind and I found myself leading the fish across the stern. The boat pitched hard, kicking up her stern and constant strain adjustment was necessary because the fish seemed to hug the bottom and did not rise to the upswing of boat and rod one little bit. John kept rowing hard to ease the pull on our lifeline and I hung on to whatever kind of fish we had.

I could not time the play but I suppose ten minutes, which seemed like half-an-hour, went by before he began to tire and presently came up in the trough of a wave on a short line. I had a vision of a light-brown back, fin and tail and profuse spotting on the body. His head was up and I increased strain to keep him coming, John shipped oars and put out the big net all in one smooth movement. We had him. I tried to look round but was instantly blinded by the rain; I heard the stretcher unshipped for use as a priest and then the thud of its application. The oars stayed inboard and we began to creep shoreward as John heaved on the straining line. When the boat grounded and he lifted the bows on to solid sand, I did look round to see, still in the net, a magnificent brown trout. Darker than he looked in the water, he was nevertheless golden and shining and vividly spotted – a little coarse in the head like all big trout in waters of that country but a very handsome fish. He had the Zulu. Six pounds, we said, and six pounds and two ounces he actually was.

We put the rod and loose gear where it would not be whisked away by the gale and using oars for rollers dragged the boat up the wet sand to the heather where in a little gully we strapped her down keel uppermost and with lacings of rope anchored to heavy stones. The day was over – in a weather sense a bad one with more of early March in it than of August. The fish were thrown hastily into the bag and we made for the shelter of the rhododendron thickets en route to the car to shed the wet clothing, get warm again and light cigarettes in comfort. This to me is always one of the blessed moments in a bad day.

As for the gamble, well, I suppose it came off. On the one side seven seatrout and a big brown, on the other such a buffeting and such an element of hazard as only the lough of the big wind – and one other that I know – can give.

XVIII

Boat Handling

A COMMENT by a gillie that to judge from the things one sees, very many aspiring lake fishers need guidance on boat handling, particularly on big lakes, has prompted what follows. I was tempted to take the line that no one should attempt to fish a big expanse of water without a skilful man at the oars or otherwise in very settled weather conditions but this, whilst it may be cold, hard sense, does not take account of all sorts of factors. Good gillies are scarce in many places and what visiting angler, unless very experienced in weather-sign reading can know what is 'settled' weather in any locality new to him? Ranging a little further afield and quite apart from the simple issue of prudence and safety – what about working the boat in all sorts of wind and wave? I have written something about methods of boat fishing in earlier books and a modicum of this will bear reiteration here – success so often depends on the approach, and that in turn always hinges on good boat management.

At the outset I had better confess, in the context of safety in boats, that I have not always practised what I now preach. Some of the risks taken in years gone by evoke shudders in retrospect. Old, cranky leaking boats, rotten oars and long crossings of very rough open water in little tubs quite unsuitable in prevailing weather were a commonplace of lough fishing in Ireland within my memory. I took the risks without appreciation of their degree. One caught fish and obtained what is nowadays called 'a kick' out of it all and long familiarity with boats in some way mitigated the hazard.

The average angler who likes lake fishing may be called on to handle a boat only while on vacation and his knowledge accumulates slowly. A gillie may be off duty for the day, considered costly, or to be totally unobtainable locally. Whatever the reason the visitor wants to go out, alone or with a companion angler, and in this situation there are some principles to be followed – varying with the water to be fished, and with a qualified estimate of weather prospects at the time of setting out.

First of all then, the water to be fished. Most small lakes are safe enough in all weathers and a wind which will result in one being 'blown off' a big water may produce no more than a brisk choppiness on a small one. Much depends on situation in relation to wind direction – a big blow on a low-lying inland loch, tree-surrounded and sheltered by low hills, may produce annoyingly gusty conditions or make it hard to pull upwind from the lee shore but it will seldom be dangerous. A similar blow in a small loch in a deep glen where the wind comes to the water between funnelling mountain walls, can produce fearful results. Two places I know, one in Scotland and the other in Ireland and described in Chapter XVII, are bedevilled by the most frightening squalls even on a day of moderate wind elsewhere. I have seen a south wind on the first and a westerly blow on the other take the tops off the waves and lift solid masses of water high into the air. This is not spray but solid water; a boat caught in one of these uncanny up-draughts has little chance. Places of this kind have an evil local reputation, fortunately, and are not likely to be visited without local guide.

It is the really big lough that presents the problem for the prudent stranger because whatever its location, shape or the direction of its long axis, a big wind will raise a big sea somewhere on it. For this reason alone it is seldom good sense to fish single-handed in the boat and never to fish lee-shore water or in places where one has to pull upwind to come home. In some lochs one can, at the worst, land and walk back but the shore can be unsuitable for beaching a boat; even in good weather on a good beach one man of average strength has his hands full to get a boat well up and safe. In such cases on soft sand the oars can be made to serve as rollers as can flat pieces of driftwood so long as the keel iron is unbroken. The boat will come up on either with a fraction of the pull. I have known my gillie and I, blown off by a totally unexpected gale, have a real struggle to get the boat up and dry and to walk back over rough ground with all the gear and a burden of concern for the safety of the boat. It is worth remembering that if you are driven aground on a shallow lee shore each wave fixes you more firmly and you can neither get off nor move the boat in to safety. True, she will not drift off but she can pound to pieces. I have seen a boat, admittedly not new, broken into pieces not more than two feet long in one night of a gale on a big Hebridean loch. She had been pulled high, dry and apparently safe the previous evening.

Even if there are two fishermen in the boat and two therefore to take an oar if need be, a day on a big lake should be planned. If neither man knows the place well there should at least be local advice, listened to with the map spread out and studied. So very much depends on the result of wind direction first and its strength second. Some winds have a trick of seeming innocent enough in the morning and of getting up later, at the turn of the tide in coastal areas, or at any time. The likelihood of this is to some degree related to season and, of course, the spring fishing suffers most from unpredictable weather switches.

The need for local guidance springs mainly from the influence, which the stranger cannot assess, of the surrounding land contours. A strong west wind on one's home loch may be all that is good and desirable – on a strange water just the reverse. With wind direction and strength appreciated, the aid of the map, and some advice from one who knows the water well, you may easily find accessible fishing ground in sheltered bays, under the lee of high shores and so on, where an ungillied boat can spend the day. I like rough weather and water but even with a well-found boat and competent help, I would rather spend a *truly* bad day trout fishing for half-pounders in some sheltered cove than taking an undoubted risk on some spray-lashed shore where the chances of salmon are good. It can happen that the boat-landing is badly placed in relation to the wind of a given day and that for this reason you cannot get afloat but this rarely occurs, landing places having been established long ago with an eye to most weather conditions. Once afloat, sheltered water should be locatable on any lake with a deeply-indented shoreline but any course which involves the rounding of long exposed promontories is to be avoided. Water tends to be shallow off these points and in a strong blow, very rough. An exhausting struggle with both fishermen taking an oar each can rob the day of its pleasure and, in ordinary times, it is simply not worth it.

Before I proceed with the matter of boat handling in a range of conditions at and on the way to the fishing ground, it might be sound to consider briefly the equipment of a boat for a day on a big loch. Check first that the oars are a pair. Where several boats are moored together it often happens that oars are changed round and the pairs mixed. Add a third oar (if you can do so without depriving another boat of its complement) because an oar can be lost and is not easy to recover unless water is calm. See that there is a stretcher or foot-rest for the oarsman; it makes for good strong pulling. If, as is all too common in fishing boats, the stretcher is missing, find a piece of scrap wood, or a stick that will fit the slots. An Irish gillie's trick is to wedge an old paint can so as to get a purchase on it with at least one foot. Next and very important in old boats, you need a baler – failing a real one the paint can aforesaid will serve. In nine cases out of ten it has to do so! Lastly make sure you have a good sound painter or mooring rope – the longer the better. This, firmly knotted across the middle of the oars, will assist to make a very useful sea-anchor should you, by any unhappy chance, be caught in a big sea unable to hold the boat's bow to wind and wave. Some fishermen, a very few, use some form of sea-anchor or drogue to retard the rate of drift while actually fishing – usually a bucket or a heavy stone on a rope long enough to trip the bottom but I do not like this trick. It disturbs the water and it can be a nuisance when a very active and fast-running fish is hooked. It has, in my view, no place in salmon fishing.

Usually rowlocks are wired down but if they are not, use some stout cord or scrap wire or uncabled painter rope and secure them. A lost rowlock is almost as serious as a lost oar on a rough day though a makeshift rig can be fitted by using a loop of rope. You drift a long way downwind however while you are rigging it up!

Drifting over fishing ground in a light breeze presents no problem except, where there are two anglers, that of keeping the boat trimmed and square. If the man who has been serving his turn at the oars stays in his rowing seat it will be found that the bow is too high for one thing, and that since it catches more wind, will swing the boat out of 'square' and for another that the two rods are too close together, wasting water and fishing uncomfortably. One man should therefore transfer at the start of a drift from the rowing position in the bow, taking one oar and one rowlock to the forward rowing place. He can then fish and with the transferred oar on the side away from the water still to be fished, take an occasional short stroke to correct the boat's course. It is not a very onerous job and in a light breeze and open water the oar will be out of water with loom in the bow and blade in the rowlock most of the time. These are the easiest and pleasantest of fishing conditions and it remains only to say that any and all use of the oars, whether to correct course while drifting or in making more extensive boat moves, should be quiet and steady. It is just as effective to row gently at moderate speed as it is to pull hard with loud-clanking rowlocks. Style too is out of place and feathering of oar blades bad – bad because wet oar blades flash.

If a light breeze and ripple are easy conditions – a very light crinkle or a calm are difficult ones. Avoidance of surge from the oncoming boat is vital and the way of creeping a boat within casting range of rising fish without putting them down is an art. Minute quiet oar dippings so that rowlocks do not jump, and good judgment on when to cease all movement and allow the boat to carry her own way are part of the art. Patience plays a big part; so too, though it is hardly within the province of the visiting angler, is the hull shape of the boat. Some boats will creep in a calm without bow-wave or very little of it and others will not.

I have not written anything in these notes regarding the use of boats with outboard motors. Their proper use on lakes is restricted to movement over long distances where the fishing ground of the day may be far from the starting point and where one may otherwise have to pull back upwind on the return. But I do know that regular motorized traffic will spoil fishing and there is perhaps an element of danger in the use of motors in rough weather. Fast-running downwind in a big following sea can be dangerous; the seas must pass the boat not travel with it. The boat should be trimmed with most of the weight in the stern half. The use of outboards for trolling is inexcusable on fly fishing water and a selfish practice which I fear seems to be growing.

Experienced boat fishermen reading this chapter will

call much of it elementary commonsense and indeed many of us observe the rules of prudence, use our weather wisdom and our boat handling skill, and take it all in our stride as part of the delight of fishing the wider waters. There are, however, plenty of anglers experienced enough in other branches of our craft, who are occasional boat fishermen only and therefore likely to gain from advice. This, if excuse is needed, is mine for inclusion of basic advice on boat matters in a book not primarily for beginners.

XIX

Salmon Problems

ONE of the brisk risers, bless them, to whom I referred in the Introduction is a close friend, a salmon fisher of some years' experience, an observant angler, and a good man with the trout. He put three questions to me reflecting problems so knotty as, at first sight, to defy constructive comment. His notes went into the general accumulation of scribblings on matters for inclusion in proposed chapters of this book and I kept looking at them and debating how to deal with them and whether I could in fact contribute anything of substance in reply. In the end I concluded that two of them, despite their varying degrees of knottiness, concerned situations of fairly common experience and therefore of wide interest, and that the third recorded an extremely odd happening which cried out for discussion in any study of fish behaviour.

Here was, in fact, the very kind of thing I had invited and, more to the point, if these questions intrigued and puzzled one angler, there must be many more with equal curiosity about similar or parallel experiences. Clearly, though I think two of the three are not answerable with any certainty I must do them all the justice that my own observations of salmon and seatrout ways makes possible. The exercise may only demonstrate how little one has really learnt in all these years of watching and pursuing the fish but it will, at all events, have the honesty of purpose that my friend and the subject deserve.

The first problem was contained in a comment: 'The behaviour pattern that really does madden me is that which arises when, in the early summer evenings, say between six and seven-thirty p.m., salmon rise literally all round one, over and under the line and very nearly between one's legs and won't look at a dam' thing – fly, shrimp, bait, prawn or worm!'

To begin with I think that this catalogue of lures is intended to emphasize the apparent futility of fishing and is not an indication that all these things have been tried on such occasions. Lacking is a guide as to the state of the river but I suspect that these are summer level if not low water conditions though I should have expected, in such case, that a big showing of fish would occur later than the hour mentioned which is no more than late afternoon in summer time. Still, it *can* happen at any time.

Now if there were many fish *and* they were not too stale, I would have thought that very careful fishing with fine tackle and very small flies would find a taker. Cast could be 1x or five-pound nylon and the two flies not bigger than tens. The line should be greased and the flies therefore fished at current speed. If the water is dead, then the flies should be drawn very slowly and smoothly the cast being directed so that the flies fish into the estimated path of fish that show. There is, of course, a right and a wrong way to fish every pool and the current acting on bank and bottom contours to a great extent dictates procedure. But fish showing in the manner suggested are likely to be moving about and it should not be hard somehow to show them the flies – not under the rod – but in attractive manner. One would increase fly size if the water is truly 'fly water' – that is to say something higher than summer level but rivers fish at different heights and there is an optimum for each.

If, as I have assumed, water is rather on the low side, and the small flies failed to produce results, the shrimp would be my next choice, fished slowly and preferably drifting, again on light tackle – indeed the shrimp cannot be fished otherwise. It is, if anything, deadlier than the fly but more trouble since it has to be mounted and I would personally prefer to begin with fly which would not spoil the water for other methods. Prawns I do not like and I think it would be in the general interest if they were not fished at all. To fish it in the conditions described would, in my opinion, be poor tactics apart from any other considerations since the chances are that if, as is likely, it failed, the pool would be the worse for it. I would have all the time in mind the fact that late afternoon is a bad time as a rule and that a pool with that kind of stock in it should be cherished for just after sunset fishing.

Bait fishing a low river is out of the question; worm fishing, where it is permitted, rather less so. I have fished worm for salmon and caught fish this way, finding it not quite as simple as it may look though not demanding high skill. Some worm fishing, for example, upstream worm for trout or seatrout in very low clear water does require a great deal of skill and a very superior knowledge of trout ways; there are probably fewer skilled performers in this method than in any other way of fishing. But skill of this order is not required in worming for salmon and I should not use it unless I wanted a fish very badly for the house. It is a dirty business as applied to what, in my old-fashioned way, I regard as a noble fish. In the case under discussion I have to assume that the pool is suitable for trundling a worm. Whether one should use it is a matter for one's own judgment; for my part I would reserve that pool for late evening fishing with the fly and if I wanted to worm it would only do so next morning.

So that is it. I know the state of affairs that my friend has in mind and I know the frustration that one can feel, but so very much depends on the characteristic behaviour of salmon in that particular river, the pool itself, the freshness or staleness of the fish – and the general air of the whole business that any advice seems too detached. I would like to decide on method and tools of attack at the waterside, watching the fish and the water and sensing the overall mood of the moment.

The second set of circumstances put forward as especially frustrating was described as 'what to me is an impossible condition – a low setting sun shining right down the length of a pool. I have never to my knowledge moved a fish in these conditions though I have seen fish rise and move of their own accord.'

No fisherman has any monopoly of helplessness in cases like this – we all share it. And it does not have to be a low or setting sun though that is possibly worst of all. Any sun shining directly downstream is bad – morning, noonday or evening, and running it a close second is a red after-glow reflected on the pool after sunset. I cannot recall a single instance of hooking either salmon or seatrout with bright or strongly-coloured light from this angle in relation to river flow. The explanation, at least with direct sunlight, would appear to be the simple one that the fish, with bright light directly in front of them, do not see anything offered to them. If the water is smooth the sun shines with little diminution in intensity all around them – if it is broken the same down-pouring of sunlight must create a dazzling world of constantly shifting flashes and glints. This latter must, of course, be true whatever the position of the sun but fish will see and take a fly in bright light, all other conditions being reasonable, so long as they are not facing into it. It is detectable in lochs that when waves are big enough to induce salmon to head into wind and wave – as they undoubtedly do in heavy water, they will not take when the wind blows out of the sun. Excessively bright sun is never good but fish will take in broken river water or in a good wave so long as in either case the sun is not 'in their eyes'. About the red after-glow I am not at all sure why it keeps fish down. It is hardly bright enough to dazzle them and my own belief has always been and remains, that there is something in the colour make-up of the red glow that destroys for the time being the fishes sensitivity to shape and tone value in the fly, and to its movement. Not all reflected glow has similar effects because I once caught a salmon on a Donegal river in the brilliant reflection of a big gorse fire on the further bank and I know of one other similar case. Admittedly two occurrences over many years do not prove anything but they do render some weak support to my idea anent the peculiar quality of a *red* light on the water.

Unless a river's course is unusually straight it should be possible to find pools around bends where the sun does not shine directly downstream. There is no point whatever in flogging a pool where it does, because it is a condition about as intractable as anything in salmon fishing. The after-glow trouble is seldom of long duration and one can afford to wait for it to fade. There are places on most rivers too, where high banks or trees will screen the water.

My friend's third question concerned curious occurrences. 'On two occasions I have watched all the occupants of a pool, salmon and seatrout, move towards the head of the pool and having congregated there go mad for about ten minutes before turning back into the body of the pool and (presumably) taking up their old positions. On neither occasion was there any apparent rise in water level nor any entry of irritant matter into the pool. No one has ever given me any plausible reason for this sort of behaviour.' I know a little more about these happenings than is contained in the précis; I also know the pool concerned, in water of an east coast Scottish river. A description here will help the overall picture and serve for comparison with the scene of another incident of the same kind which I saw and will set out as it happened.

The pool in Scotland is a long one with an equally long shallow upstream of it. The neck of very fast water is narrow and of moderate depth. Below it the pool widens to thirty yards or so, fringed with willows and mixed scrub on the shallow side and steep-to opposite where it runs deep against timber shoring. The whole pool down to a very nice tail is about one hundred yards long with reasonable current all the way: it holds a big stock of salmon and seatrout, some of both very big fish. There is a small town a mile or two upstream and the sea is not far away. My information is that on both occasions the river was at normal summer level neither excessively low nor high. Both occurrences were in July *but a year or two apart*, and both about noon in hot weather. These were the scene and the conditions: the bones of what happened are in the words I have quoted but the verbal account given to me suggested a far greater element of excitement among the fish. Apparently they did go crazy, darting to and fro, turning and twisting and gyrating in a manner which held my friend spellbound. They might well have been in the throes of dying from some irritant poison but of course were not and when the mass hysteria had evaporated went quietly back from the head water into the pool below.

Before attempting to speculate – and that is all that I can do – about this very curious behaviour, I will deal with my own similar experience on a North Devon river. The pool concerned has marked similarities with that on the Scottish river – not quite so wide perhaps in its hundred-yard length but wider at the head where there is a long rough stickle. Again there is moderate depth and reasonable current and the pool is a great stronghold of both salmon and seatrout, carrying salmon from March and biggish seatrout from May onwards.

One dull hot August day with the river fairly low I arrived at the head of the pool to fish it down with greased line and small fly. I was on the wrong bank which is high and was passing on to wade across higher upstream when, walking well back from the edge of the high bank, I saw a white flash in the strip of water not hidden from my view by the bank – and then another. I propped the rod against a bush, shed net and bag and crept softly across to the cover of a bush. There I could see that the area where the entering current was losing its velocity was full of fish. Salmon and possibly big seatrout were rushing in all directions in water about three feet deep, many turning on their sides and showing bright flashes of near-silver as they did so. The fish seemed to come to the edge of the broken water and turn there to circle again. The entire width of the river was involved and the fish were impossible to count – there might have been twenty, thirty or forty. Twenty years' experience of that pool all through the seasons enables me to estimate its average stock which is about thirty salmon and perhaps ten seatrout of three pounds and more. I think that the entire stock was milling round like demented fish as I watched them. There may easily have been more salmon than I could see without moving from cover, and there may have been dark fish and seatrout without the brilliant shine of their fresher fellows, and therefore less visible. All the fish were lunging, turning and twisting close to the bottom – here of fairly large stones and none broke the surface.

I watched for perhaps five minutes and then the activity ceased. How long it had lasted altogether I have no means of knowing and I did not *see* – as my friend did in Scotland – that the fish dropped back into the pool. They probably did but the river bed is dark-coloured and there are clumps of trailing weed so that fish can be there unseen – invisible as soon as they cease to turn. My first thought, as I backed away from the edge, was that there had been a rainstorm miles away, high on the moor where the river gets its water and that the fish felt the first fraction of a creeping rise. Against that, there had not been any long drought and a little fresh water would not normally create so much excitement. That was simply tested anyhow and I went up to the ford to inspect the dry stones on the rim of the water. The little lapping waves there had the concavity of water tending to fall rather than rise, so it was not that.

The turning-over movement of the fish bothered me. It was precisely as if they were rubbing sides on the stones and so getting rid of parasites. If so what kind? None were fresh enough still to carry sea lice, the freshwater louse is not common enough in that river to worry the fish and gill maggots, which doubtless many of the fish had, could not be removed in that manner. And anyhow, what would prompt a kind of communal delousing in which all the fish had a similar impulse? Deliberation on this occupied a long drive home that evening and I still think about it. The movement along the bottom, the fish turned on its side and body arched towards the stones, is exactly the action of a salmon trying to dislodge a hook. I know because I have twice allowed a hooked fish liberty to do this in a place where I could see every move. I could see the fly protruding from the scissors and the fish trying to scrape it away on the pebbles. Both fish had been reasonably played out and had been brought into two feet of clear water. I draw no conclusion from the fact that fish trying to shed a hook behaved precisely as the crazy fish in the pool – I merely think it interesting and very curious indeed.

Before I return to the Scottish incidents and their consideration, I ought to say of my day in Devon that no rise of water ever developed and that my efforts to fish the water after crossing to the better fishing side were fruitless. The pool, throughout its length, was dead – sleeping in the heat of a sultry day. One final point – it was half-an-hour before noon, sun time, when the fish were active.

There are remarkable similarities between my friend's two experiences and my single one. The constants in all three were weather, water level and time of day – with an approximation in season – July in two cases and August in the third. Of these, I regard the high air temperature and relatively low water as of most significance. In the account of the experiences in Scotland there is a reference to there being no 'apparent' rise in water level, but I know that no rise materialized later and I therefore accept that no 'creeping' rise, such as I at first suspected in Devon, was present when the fish were behaving so strangely. I doubt if it would have led me anywhere but I would like to have seen all three cases with my own eyes to determine by the same standard what *proportion* of the activity took the form of turning and rubbing on the bottom. All the fish in the Devon river pool did this – and repeatedly.

In my experience the only perfectly natural cause of general excitement on this scale among fish confined in a pool is an approaching rise of water. Even then it requires in most cases the stimulus of the first minute increase in level and is only seen when, broadly speaking, a drought is on the point of breaking and the river extremely low, which it was not in any of these cases. An absolutely natural cause is out of the question and we must look for an exceptional and artificial stimulus, operating in the combination of high air temperature and low river. Together they could, and probably did, produce a shortage of dissolved oxygen in the body of the pool. There *is* a flow through both pools, but in the centre third of each there is a body of nearly dead water and it was in both cases from this area that the fish came to the more highly oxygenated water at the head.

There have been suspicions of slight pollution in both rivers at times, both having towns on their banks some little way upstream from the pools concerned. I suggest that all these cases *may* have been due to pollution by some irritant effective only when insufficiently dispersed by low water, or by further reducing oxygen content on days when weather conditions produced exceptional loss of oxygen from the water. The dead water could conceivably become intolerable to fish and a move up to the more highly aerated head water would be natural. Pollution of this character would, on the other hand, be unlikely to be of such short duration. Again, would shortage of oxygen pure and simple produce a frenzy in the fish and would it not be extremely odd if two rivers five hundred miles apart both suffered from the accidental release of some irritant polluting substance? We do not, in either case and very unfortunately, know whether fish in other pools higher up were affected.

As will be evident – and I hope that you will not have felt you have been too long at this inquest – I have been intensely interested in these strange manifestations of salmon behaviour. If they spring from natural causes they are outside my experience of looking at salmon and I would dearly love to know more about them. If, however, as I personally suspect, they have some chemical stimuli, one would be equally anxious to know more – for obvious reasons. It may be that some little light has been shed by this attempt to argue the matter out and to provide my questioner with an explanation at least plausible.

XX

Lough of the Crannog

IF YOU fish in Ireland, or in parts of the Highlands and Islands, you will add to your pleasure if you can take some little interest in archaeology. Naturalist or student of nature you *must* be wherever you fish and you cannot be a fisherman in the full sense of the word unless you are. But some small understanding of archaeology is *not* necessary – it merely adds interest to your doings in country where there are prehistoric hut circles, lake-dwellings and other traces of ancient peoples whose pre-occupation like yours was the taking of fish. The direct link with your fishing is, of course, tenuous but it is both truthful and not stretching the point too far to say that over the years some hundreds of trout and seatrout must have gone into my bag because of recognition of lake-dwelling sites and their underwater features.

Actually it bears the name of an Irish saint, this lough of my story and I had no idea until I saw it that it contains the best example of an Iron Age artificial island, a crannog, of the many I know. The country in which it lies is dotted with loughs large and small, bog-fringed and peaty for the most part. Some are unapproachable, the margins floating mats of vegetation which slowly sink underfoot; one makes half-a-dozen casts to rising trout and must then flee to firmer ground. Some pools on the high ground are hard-bottomed and weedless and here and in the peat loughs the trout are small and dark: any team of small flies will catch them three at a time. Big trout are those of a pound or over but in a few places there are really big trout; I know of a worm-fishing small boy who produced from a piece of sacking in the village one evening, three lovely trout each over four pounds.

I heard about the lough named after the saint, during a talk about big trout. It was the property of the owner of a pub in a nearby village, and so private that even gentlemen in dog-collars were not permitted to fish. The owner, I was informed, was 'a cross man – as cross as a bagful of weasels'. He had stocked the lough, an unheard of thing in those parts, years before and there it was: he did not fish and he did not allow anyone else to fish. Legend had it, of course, that there were enormous trout there. I proposed to make friends with the cross one and was strenuously discouraged in any such venture: I would return, I was assured, with a very large flea in my ear. But I called at the hotel, where the purchase over the bar of enough wine of the country to float a curragh, and the undertaking to do the landlord some trifling service in England, gained me the freedom of the lough for one day. I did not find him a cross man at all; in fact towards the end of the evening he actually produced 'me private bottle' containing amber liquid fifteen years old and we became distinctly friendly. I was bidden to take the boat – 'She's a grand boat, the best ever to come out of Rathmullen.'

My gillie John knew where the lough lay in a jumble of small rough hills out on the moor but I was totally unprepared for what I saw when we reached it through thickets of scrub hazel curtained with hanging lichen and acres of tumbled boulders hidden in old rank heather and bracken. Steep-sided, the lough occupied a small hanging valley and covered perhaps ten acres. An ancient glacial outflow was marked by a low mounded ridge across the only gap in the skyline when I looked across the lough from the water's edge and I could see the moor far beyond and below. We had intersected the shore almost at the north end and here it was low and reedy. A deeply-indented bay lay to the right and another, apparently filled with lily-pads, ran in towards the overflow gap. One could fish the shore at the outflow but nowhere else. Nine-tenths of the shoreline was under boulder-filled undergrowth and very steep, or was forty-foot cliff as smooth and vertical as the wall of a house. Immediately before me, forty yards from shore and across the mouth of another bay was a small oval island, tree-clad and scrub-shrouded, which looked like a typical crannog.

It was a nice evening in May, clear and warm but the breeze was all over the place bringing a ruffle from one quarter and then another. It did not look promising though I had been warned that it was useless to fish before dark when the absence of breeze would matter least; the thought of the salmon I had had that morning from the big lough was sustaining. Odd trout were showing far out presumably taking sedges blown out from the shore. The water was extraordinarily clear with a greenish tinge unusual in that part of the country.

We found in the reeds 'the best ever to come out of Rathmullen'. She was resting on the bottom with only the gunwales above water and without signs of oars or baler. Search of the undergrowth produced two odd oars minus leathers and a rusty paint can which would serve to bale and eventually a very rotten boat was floating. We conceded that at some time she had been a good boat. Now she leaked like a basket and if one allowed her to fill again she would sink like a stone. The worst leaks were plugged with pieces of rotten mooring rope and keeping booted feet on main frame members we pushed off in the certain knowledge that a foot on unsupported planking would go straight through.

Without knowledge of the water, I put up a test cast with small Mallard and Claret on the tail, Corncrake Sedge on the dropper and Butcher on the bob, and by the island we started to fish. There was insufficient breeze for a drift and I threw a long line from a quietly moved boat. Occasional fish showed but none within reach and I fished blind towards the overhanging branches and trails of honeysuckle on the island where I presently had a deep quiet rise. I played the fish while John baled out some of the rising water and he had done so and taken to the oars again before the fish was ready for the net. I have never seen a better trout for shape, condition and colour. It was a short deep fish – sparsely spotted red and black, of clear golden yellow on sides and belly and an obvious three-pounder. This was good.

Before moving elsewhere I had a closer look at the island and there, plain to be seen going down ten feet into the clear water were the great timber piles, oak probably, that held the island together. The Iron Age people who made these crannogs nearly 3,000 years ago, often used walls of piles to contain the brushwood, rocks and earth piled inside them but usually the piles have sunk or rotted or splayed out and now lie flat and the island sinks lower as the ages pass. This was a good example of the uncausewayed lake-dwelling: I could find no trace of the underwater track of stones connecting crannog with shore that form a feature in some loughs and provide, where they are long enough (some may be 100 yards) good shallow fishing ground. Some of these causeways have twists and breaks in them, clearly to mislead an enemy who might identify the shoreward end and seek to cross. Seatrout and trout find these artificial reefs much to their liking and I always fish over and around them when wind and wave will permit. I know one at least where salmon lie, the irregular stony track, now carrying two feet or more of water always holding a few fish. The archaeological diversion thus I hope justified, I must return to the events of the fishing I am describing.

The area round the island did not produce any more rises and I baled furiously while John pulled across to the bay of the lily-pads. The light was going and with sunset we lost the wind: the cliffs and high ground above them cut off the light from most of the lough and left it a still pit of gathering blue shadows. It had had a curious sombre air in full light and now out on it in a sinking boat in the nearing dark I decided, childishly, that I did not really like it, in spite of the beauty of that trout drifting round in the water at my feet.

But I went on working the flies slowly and deeply in the calm and expecting some sort of general night rise to begin at any time. On the edge of the lily-bed, I had another take – a fish of about two pounds this time but light-coloured and bright like the first. After that there was no sign of life and the lough was utterly still. We put the boat's nose on the beach behind the reed bed, baled again and considered. 'If there *are* big trout here,'

I said to John, 'they should move in the next half-hour and we should give them a try. You watch the water level and I'll fish a really big deep fly.' It seemed reasonable enough and I mounted a 9/5 cast and a single big fly – a number six Dusty Miller. In retrospect it seems a strange choice of pattern in view of the absence of any light at all and I do not now recall my thinking when I selected the fly. I do know, however, that it was one of the queer variety of Dusty that has a lime-green hackle in place of the pinky-flame of the regular dressing.

We moved out towards the centre of the lough and into, presumably, deep water. There on the edge of the hill's inky shadow I began to work the big fly with slow line-fingering. The boat would go quietly with a touch of the oars for so long and then become loggy as she filled and we would have to bale and move on to an undisturbed area. We could never bale her out and we counted her dry with five or six inches of water round our feet – baling at any lower level the paint can banged on the ribs and in that still pool we needed utter quiet. In that quiet we could hear that most elusive sound of early summer night in that country, the croak of the flighting woodcock among the taller trees, and the gabble of duck off the nest to feed and splash along the margins.

I was thinking that we had had enough of this place, becoming increasingly eerie and depressing, and of the menace of the sinking boat, when the rod-top slammed down to a solid pull and I was into what was obviously a heavy fish. I got rod and myself into some sort of conventional fishing attitude. I had not with such slow fishing expected any sort of surface take and I had been fishing by feel alone but John, who had not realized how deep the fly was, had been watching and now said he had seen the fish's body and tail as it took the fly. I had not. 'Like a salmon', 'father and mother of all the trout', 'begob he had a tail like a shovel' were expressions I heard in the next few minutes while the fish cruised deep with the reel ticking slowly and John turned the boat with one oar and baled furiously with the other hand. The fish pulled hard and evenly and was about as unyielding as a salmon moving at a speed of his own choosing. The ten-foot rod was hard bent and the line entered the water an oar's length from the boat. We moved away a little but the fish came with us. He kept making little runs but never away from us – always down and deeper. I tried easing the strain to see if less pull on him would induce him to come up so that I could shorten line but this only encouraged him to take more. Nothing I could do with the rod was effective and so much line had gone that we dare not again try pulling away to get side strain, because if he then made a run, we were done. I did not know how much backing there was on the reel. I had thirty yards of running line and some backing and I regularly used it for seatrout fishing where I reckon one is safe with fifty or sixty yards all told. Most of the dressed line was now out.

Any thrill that this battle with a big fish possessed lay in its demonstrations of sheer strength and obstinacy on the part of the fish and however protracted an account of it may be, it will reflect only that. At the end of something like fifteen minutes of circling and holding on – and baling – the line was straight up and down and the fish was right under the rod-tip with thirty yards of line and some backing out. I had a mind's-eye picture of a very big trout indeed cruising in the darkness about a hundred feet under that rotten boat. The thought was all but oppressive; he seemed tireless.

How long he could have circled in sullen strength I shall never know because without variation in the strain and without warning he was gone. Cast and fly were completely in order. We baled the pride of Rathmullen for the last time and paddled her back to her place among the reeds where, as far as I know, she still lies.

I still wonder, sometimes, if I could have handled that fish differently. My handicap was lack of line. One of my salmon reels with a hundred and fifty yards on it would have enabled me, without risk, to make him run. . . .

XXI

Injured Salmon

IT IS widely believed in the fishing circles in which I move that in recent years we are seeing more and more salmon carrying scars of earlier wounds, or newer injuries in various stages of healing. The explanation that predators are more plentiful than in former times is rather too simple but no reasonable alternative seems to present itself.

At one stage, I thought that I had detected greater evidence of wounded fish in some coastal areas than others. For example, in the wild, remote river systems of the Atlantic seaboard in Scotland and in north-western Ireland where seals, otters, killer whales, porpoises and all manner of predators are present in inshore waters in relatively large numbers, I thought I saw the results in a high proportion of scarred salmon. But after some years of studying such areas in comparison with others less predator-haunted – like some in the south of England, for example, and in the West Country, I am not so sure. I am beginning to wonder whether, in fact, the preponderance of injuries are not sustained far out in the open sea; pointing this way is the fact that the majority of wounds are pretty well-healed. Some have obviously originated a long time ago, probably in the smolt stage and have all but disappeared, others at some adult stage. It was natural enough to assume that the greatest damage among salmon would occur where both prey and predator were densest – in the estuaries or river approaches. I now think that this is probably true only for shoaling smolts at migration and possibly for finnock. We now know that even greater densities of salmon are found in far-off feeding grounds of which Greenland is one.

Of course, some damage will occur among fish approaching their rivers and shoaling in the estuaries and probably these injuries – not very many – which can be ascribed with accuracy to otter bites would occur here. Seals are abundant and active among salmon close in-shore but again one seldom sees fish with fresh wounds, i.e. those clearly sustained immediately prior to entry of the river. It has been said of one east coast fishing that six per cent of captured fish exhibit seal damage. I would like to know, as an indication of how many are injured but survive where fish are shoaled in numbers in open water, what proportion of scarred fish come to hand in Greenland nets.

Prompting interest in the entire question was and is the fact that injured fish are readiest to take. I personally find that on days of poor conditions when fish are 'off' so that one records one fish for the day – that single fish so often bears an injury. For the most part they are flesh wounds rather than structural damage though I recall one very active salmon which was lacking the upper lobe of the tail and another in which a half-moon cleanly bitten out of the back had taken the dorsal fin and supporting tissue. There has never been in my experience anything in the play or strength of a fish to indicate injury which has always been unsuspected until the fish could be seen close to boat or bank, or otherwise at short range.

What makes a wounded fish so (relatively) willing to take fly or bait? Puzzling over this question for many years now, I am still without a satisfactory answer. Pain, in the sense in which we know it, is not a factor but irritation arising from something else – disturbance of the nervous system perhaps – may be. The direct effect of healing or partially-healed tissues, or wholly healed but misshapen body surface, is another unknown. Or may it be that a salmon which has escaped, albeit damaged, from some terrifying enemy attack is the more alert for anything strange, moving and unwelcome that comes within its ken and is at the same time of a size to be vanquished? I lean towards this hypothesis though it does not quite square with my belief that wild creatures – whether fish, flesh or fowl – living in a milieu which bristles with enemies and consequent dangers, are ever other than completely alert, injured or not.

What is special about the wounded fish may be that the defensive/irritant mechanism is triggered the more readily – to express itself in pursuit of and seizure of the passing object. I do not know and am unlikely ever to know but, like most fishermen with experience of injured salmon, I find these questions of constant interest.

The nature and origin of wounds is a minor study also of considerable interest. It rounds out some part of your appreciation of the life of the fish and never to provide vivid evidence of the hazards besetting salmon from smolt migration, through sea life and during return to the headwaters of the river. As often as not, a captured fish after suitable application of the priest is laid in the grass or in the boat with a hasty judgment of its weight, and one is fishing again. Good fishing sense too, because the hooking of one fish may be the indication of a general move, and that move may not last long. But later there should be a good look at both flanks and the ventral area of the fish. The commonest marks on a fresh fish are the slight blood suffusions from the attention of lice but you may also find, usually on the throat, near the roots of the pectoral fins or on the belly, the circular rasp marks of the sea lamprey. I have often seen raw contused areas less regular in outline than the lamprey marks and these I call rubbing marks resulting from sustained contact with rough stones in some lie. But, against this, we know that salmon make use of the anal fin to judge their level above the gravel in spawning beds and it is difficult to believe that the fin is not used at other times in this manner. If so, no fish would be so close to the bottom as to rub itself raw but the rawnesses are there and are not obvious lamprey wounds.

About lampreys incidentally – sea lampreys that is, how many fishermen have seen them spawning in the shallows of summer streams far inland? I have often watched them at work in North Devon rivers. They seem to come in during March, April and May with the spring salmon and according to some authorities spawn at once but I have never seen them spawning before June and sometimes later. Eighteen inches to two-and-a-half feet long they are not pretty – dirty grey and yellow, blotched and spotted with brown – they arouse all the revulsion one has for the hag fish or any other boring parasite. But they work hard to excavate the spawning bed, moving the gravel and fairly large stones, by attaching the suctorial disc and tugging the stones away. Both male and female will attach themselves to a particularly large stone and move it by joint effort. The result is a trough-shaped depression in which the ova is deposited and fertilized. The survival rate of the spawned-out adults is low but there is a downstream migration so that some may get back to sea. And from lamprey-spawning back to salmon and perhaps the digression was warranted since assuredly some of these wriggling horrors do hitch-hike up-river on incoming salmon. I have only once actually seen one attached to a fish; the fish jumped while I was playing it and the the lamprey, a small one, was plain to be seen. It had disappeared when the fish came to the net.

The otter kills only a percentage of the salmon he pursues and in my opinion he either secures the fish or loses it altogether by failing to grasp it so that very few fish indeed carry otter bites. Some small radius 'half-moon' wounds on salmon nevertheless look like the bite of a relatively sharp-snouted animal and they are certainly too narrow to be seal bites. These are unmistakable – usually to the rear of the dorsal fin, often just above the wrist of the tail and look for all the world as though the pursuer had grabbed and all but missed contact. I have often seen deep scars where segments of flesh have been bitten or torn clean away – one such wound I have already mentioned; it actually exposed the backbone of the victim which was otherwise normal, strong and lively.

On the evidence of damaged fish alone I feel that seals take many more salmon than is generally believed though I do not know enough to distribute blame fairly between common and Atlantic seals in this respect.

If a cormorant, that dark arch-villain, strikes and

misses, the evidence does not long remain since it usually consists of rough slashes of displaced scales on both sides of the escaping fish but these marks are not uncommonly seen in fresh finnock and small seatrout. There may also be damaged tails and torn fins from the same cause. Cormorants will, however, tackle quite large fish successfully. I recall the outcome of a war waged on a particularly persistent cormorant which made daily visits to a trout lake in Anglesey in which I was interested. When this bird was successfully ambushed – and it took a great deal of doing – he had inside him, among small stuff, one trout of two pounds, and the remains of another of similar size. It is astonishing to see what the 'skart' can swallow – watch him shake down a big flounder!

Some salmon wounds are beyond identification. If they are bites the evidence is on both sides of the fish. A one-sided tear may be a gaff mark though these are rare and happily becoming rarer with the disappearance of the gaff in favour of net or tailer. A single clean cut on one flank, or indeed on any part of the fish, may be a propeller slash and I have seen a number which I felt arose unmistakably from this cause. Power boats in estuaries where salmon are thick must occasionally run among the fish and there is some evidence that salmon are attracted to, rather than repelled by, propeller turbulence and wake.

There are plenty of stories of fish jumping at obstacles and injuring themselves by striking rocks and most of them in my view have little substance. So long as there is a cushion of water the chance of injury is small and fish seldom leap on to dry rocks. I have only once seen it and minute examination of that fish failed to reveal any bruising or contusion though the stones were rough and dry and the fish landed with a thump from a height of about four feet. I think the blow is glancing and seldom direct. I did, on one occasion, see two fish both exhibiting triangular deep punctures on the shoulders, looking as though they had fallen on the pointed end of a fencing post. This was a real puzzle until I discovered that their captor, a very excited – and fortunate – beginner, having failed to get them into a trout landing-net had manœuvred them into a little rock-studded pool and there fallen upon them! The truth about these mysterious wounds emerged when I asked him how, on a fine day, he became so very wet. . . .

XXII

Otters

LIKE most seatrout fishermen I have many times said rude words when the pool selected for my dusk operation has been ruined for me by the appearance of an otter or otters. But, momentary irritation aside, I have never grudged the otter his fish and what he takes in good healthy fish is more than made up for by the pleasure which study of a most attractive animal can provide.

I hear with sadness nowadays that the otter is declining – that one of our oldest Britons is no longer seen in so many of his one-time haunts; my publisher would not pass words to express my disgust and indignation and perhaps it is more constructive to take some active part, however small, in conservation than to fulminate. I would not be without the otter. He takes the fish which are his right and to that extent and only that, we fishermen are the losers. Are we so careful and unselfish in taking fish for sport that we can blame the otter for taking the food to sustain his life?

After much fishing and the opportunities of watching otters hunting and playing it has brought, I am not quite sure that the animal is the dread predator, spreading panic among salmon and seatrout, that the effect of his presence in an evening pool would appear to indicate. On such occasions he certainly puts fish down and off the take for some time, but there are circumstances in which fish will ignore the otter.

I saw it happen recently. The area is about as remote as one can find; the place the head of a narrow sea-arm at the point of entry of a small river. Here the sea shallows gently over sand, gravel and weed up to the stony tail of the river's sea-pool. Time was mid-day and tide approaching full. I was heading for a salmon pool upstream and paused at the sea-pool to look seaward for evidence of salmon coming in on the tide.

The fish were there all right – two shoals of perhaps fifty fish each cruising aimlessly on the shallows a hundred yards away. They were awash, back fins, tails and backs showing continuously: the river was low and they could not run and would go out with the ebb tide. Not more than twenty-five yards from the slowly moving masses of fish were four otters – one adult which I took to be the bitch and three rather more than half-grown cubs. The fish were in about three feet of water and the otters loosely-grouped closer inshore in eighteen inches or two feet.

I watched them for half-an-hour when the attraction of my salmon pool became too much for me. They were seldom completely submerged. They were taking something which I took to be shore crabs or some kind of shell-fish – mussels perhaps – from the bottom, occasionally by diving but more often by simply 'up-ending' so that only the rumps – one big and three small – were showing above the surface. Twice the bitch paused to stand erect in the shallow water to stare intently in my direction; I was then within fifty yards. Salmon splashed ten yards from her but she went back to whatever she was doing.

The otters ignored the salmon and the salmon the otters. Before now I have seen two adult otters and a seal working a shoal of salmon at this place so that it is by no means a question of the otter requiring his fish to be in a confined space such as a river pool.

No – I suggest that there are things in the otters' diet which, on occasion, he likes better than fish, and that in the absence of communicable menacing intent, his presence in open water does not disturb salmon.

Otters are great travellers, using age-old trackways over moorland and mountains, across lakes and along the shores of the sea – the inclination to cover long distances less marked, it seems, in the cultivated and coarse fish populated south. I know one cross-country otter track in Ireland that has been in continuous use in my own observation for forty-four years. It was first shown me by a man in his eighties who said it was old when he was a boy. It is a route across the reedy thickets, headlands and inlets of a big lough and there will be no morning when you do not find it has been used during the previous night.

Looking at it, deep-worn among the tussocks, I have wondered how old it is, how many generations of otters have used it. Thinking of it sometimes in the warm tranquillity of an English country cottage, but knowing that far place so very well, I sense the fury of a winter gale on the lough, the howl of the north wind from the Inishowen mountains, lash of rain and seethe of rollers in the reeds and the sleek otter shapes flowing about their business in the hissing dark.

I have always been surprised to find how many fishermen have never seen an otter. I suppose the thing is governed by the element of chance which dictates that not having seen an acquaintance for ten years one meets him thrice in one week. I seem to be always seeing otters or, failing that, signs of them. Recollections crowd in but, of course, most sightings are forgotten. Some however are unforgettable.

On a tributary of a Loch Lomond feeder river, I once came through undergrowth to a deep round pool with high red clay banks. One bank on my side, as I knew, contained a regular sliding place but I had never seen it in use. Now an adult and two cubs were using the smooth slippery clay with obvious enjoyment, slithering from the lip of the bank to the water's edge, out of the shingle at the tail and back up the bank over and over again. I withdrew undetected.

At the base of another high bank on a big pool on a famous West Country river, is a little rushy platform and down here, seatrout fishing at night, I used to wait for the light to go before starting to fish. It was a good place well below the skyline with tall bracken for background. I was still and because even the glow of a cigarette will put seatrout off at the critical time when they are moving from daytime lies into night-feeding places, I was not smoking. Perhaps the only scent was that of the river on my waders because between my feet presently appeared the sleek head and the forefeet of an otter. Body submerged, and muzzle not six inches from my brogue he – or maybe it was she – rested motionless for ten or fifteen seconds and then with a little snort was gone.

One day, traversing one of those chains of lochs so typical of the Outer Hebrides, on the way to a day among the salmon in another loch, I was threading my way, outboard at half-throttle, through a maze of small heather-clad islands. One of them is a regular haunt of otters and I often landed there to inspect the feeding place. There, as usual, was a flat green spot well above water level, where the otters brought the catch to be eaten. Scales, earbones and such oddments, sometimes a whole backbone, as had been left by the hooded crows in cleaning up the otters' table, gave a good indication of what the animals were taking – salmon, seatrout or trout. Opposite the island, on the western shore, was a deeply-indented bay, very shallow indeed and never fished or disturbed for that reason. And from out of this bay as I pushed off from the island, came a swimming otter. It ignored the clatter of the motor and swam steadily athwart the boat's course. At oar's length away it dived to reappear on the other side of the boat heading unconcernedly and unhurriedly for the island. These otters do not see many humans and this one was, I should say, a two-thirds' grown cub.

Three times its size was a very big otter I met very early on a June morning – also in the Hebrides. I had gone out soon after full light to fish a certain pool a few hundred yards below the river's outfall from the loch and I was fishing the current at its head when, looking downstream as I fished, I saw the rushes of the bank part at the pool's tail and an otter's head and shoulders appear. The animal slid into the pool and for three or four minutes I watched a chain of bubbles circling the pool. Fish did not show, either in panic or otherwise; the bubbles ceased and the pool was still. I do not really think that otter was hunting – I think he was returning from hunting because presently a disturbance in the broken water of the shallow above showed that he had left the pool. There were many salmon and seatrout there I knew.

I followed the otter upstream, marking him by the disturbances in the thin water – disturbances such as one sees a running salmon make in such places and shortly, by cutting off a big bend, I was able to get ahead of him and to await his coming on a footbridge just below the loch outfall. Here, strangely, the otter paused and lay for a time – perhaps a full minute, in eighteen inches of clear water among the stones under the bridge. He *seemed* to be clinging to the bottom and apart from very minor water distortion I could see every detail of that admirable body.

We parted company when he went upstream again, up the low fall into the loch to emerge in a few yards on a boulder in the early sunlight of a lovely morning. Fishless, perhaps because of him, I could only stand and admire while he groomed the guard hair of that wonderful coat. After all, he belonged there – I was the intruder.

XXIII

The Yellow Stone

SINCE I was a boy and the salmon was at once the abiding passion of my life and the greatest mystery of all to me, I have wondered and I still wonder, why the fish in loughs are so strongly attracted to lies by isolated rocks. The decided current of a river acting on and around a rock in that medium will create a micro-climate and a lie – so much is readily understandable. Its counterpart, probably bulkier and rising sharply from the lough floor, often in relatively deep water, may provide the element of cover but it is difficult to see what more.

Yet these isolated rock formations, great boulders or segments of reef do gather fish to themselves and there is a scattering of them in most rough country waters in Scotland, the islands and Ireland – some of them famous and the focus of concentrated fishing. Mostly the named ones bear descriptive labels. I know several 'Salmon' rocks, a 'Priest's' rock, a 'Pulpit' rock, a 'Kitchen' rock, a 'Red', 'Black' and 'White' rock and many more. They differ in size, character, colour and in fishability in any given strength and direction of wind. One can find a limited similarity rock to rock in one group – that composed of rocks lying in shallow water. Here the fishing rock can simply be a convenient mark on which to align the drift of a boat and it does not necessarily follow that the stone itself has any special virtue as a lie. It is merely the visible indication of a shallow area which holds fish. Any fish caught within say thirty yards of the rock is said to have been caught 'at the rock', and so it gains in repute until it comes to possess some kind of imaginary magic.

You and your gillie will look at it through narrowed eyes from afar when salmon are hard to come by and one of you will say, 'There's a nice wave on the Priest's Rock.' Presently you will shift your ground so that you will fish the rock though in your heart you know that when they are 'off' they are as down at the rock as anywhere else.

I said that your true isolated rock rising out of deep water could only qualify as a holding-place on the score of providing some sort of cover, but its underwater contours are hidden and there may be ledges and irregularities quite unsuspected and offering resting-places for salmon. There may be other hidden virtues in such a place – perhaps such a prominent feature of the lake floor comes up in particularly featureless sub-terrain – or perhaps it acts on underwater currents as it does on surface wind and wave drive. For my part I am content that there are days and occasions when salmon sail flywards from the shadowy depths beside the rock and my faith in the fishing mark finds new justification.

However cynical you may be in a non-taking period, these places do lure you on and the Yellow Stone is no exception; indeed it gains in allure because it is hard to reach. It serves no purpose to tell you exactly where the loch is since it is a private fishing: for my story it is enough to say it lies in a Hebridean waste of mountain and moor – a big sheet of water in a country that seems as much water as firm ground. Here on a bright day of blue sky, innumerable lochs large and small, round and irregular, reedy and rocky, glitter like jewels in a setting of bronze and gold and chocolate of peat, and beyond it all the roaring Atlantic – rushing green over white shell-sand beaches. Without sun the lochs are cold silver or pools of darkness in that most desolate of all landscapes and the melancholy of the Isles comes quickly to the sensitive.

The Yellow Stone lies a hundred yards off the north shore of a low headland, itself at the north end of the big loch. To reach it is a major exercise: one walks, climbs and stumbles over three miles of bad going – a large part of it that most trying of ground, small round boulders hidden in thick grass and heather – to reach the boat. Once afloat there is four miles to go on the outboard motor and because the Yellow Stone fishes best in a north wind that means, in a salmon-suiting wave, four miles of what can be a big head sea. In fact, from a wave-raising point of view the area of the Stone can be worked with an east wind but it fishes badly, like most places, with air from this quarter. So on the rare occasions when we go, we select north wind days, trim the boat carefully and pray that the trough of a wave will not drop us on one of the numberless jagged rocks that litter the channels through the little islands. It can be an interesting passage. Knowing navigation makes use of the fact that some of the islands lie across the north wind and give short interludes of quieter water. Their western and eastern ends, where they terminate in tumbles of boulders, sometimes yield a fish where the wild surge of waves across the point eases slightly as it feels the island's lee. This fishing is for the homeward passage if one has not stayed too long on the distant fishing ground.

We picked a June day, Angus the keeper and I, with half a gale blowing from the north. The light was strong and the sky as hard as a northerly sky can be but you cannot have everything, and our estimate was that across the wide bay to the Stone there would be a strong enough push of wind to build a fishing wave. I think on balance I would rather have a good brisk rolling wave from a wind in an indifferent quarter, than too little wave from a good direction.

Very few river systems in the Long Island have spring runs of salmon and the great bulk of the fishing depends on summer fish. June is rather early but at the time I have in mind we had had very high water and fish should be in the big loch. En route to the fishing ground we were reassured by the sight of two plungers, as it happened too far off to mark accurately.

Angus put the tiller over and made a small detour over the approximate location of one of the fish and throttled back while I trailed a long line and a fly for a few minutes as we passed. Either it was a remarkably accurate piece of marking on his part or we were simply lucky because almost at once I was able to say 'There he is' and we both watched, way still on the boat, while a fish swam vigorously after the fly – back and fin and tail tip showing in the wave. He took hard and went down with the fly – a No. 6 Silver Grey and we were 'in him'; Angus cut the motor and we had five minutes of lurching and slopping in a big wave before the net went under a fresh eight-pounder. Foundation for the day!

In some places and conditions, salmon will take a trailed fly very well and there are days when if you elected to do so you could catch more this way than fishing the fly from a drifting boat. I have little time for trolling with a bait but it does no harm to trail a fly in passing from drift to drift. Usually the fish take with a bang but one that follows on the surface can be a problem. He may turn away if the speed of the fly is changed in any way – he may lose interest if it is maintained. One takes one's chances. I have often hooked fish that followed, without coming really close to the fly, by quickly pulling off a foot or so of line and allowing the fly to drop back so that the fish so to speak overran it.

An otter regarded us calmly from an emerald-green patch at the back of the beach as we rounded the big headland into the bay of the Stone and Angus stowed the outboard and took to the oars to give us three hundred yards of quiet approach to our fishing. A diver flew high overhead with a quick kakkering cry of its kind. In the bay the wind was strong and even and blowing fairly on to the beach of the headland – here a lee shore. Half-a-mile away to the north-west the shoreline dipped where the outflowing river left the loch. Fish entering this way have the habit of resting in this bay and especially in the area of the Stone, which today with the loch two feet above normal, was just awash. It was mid-afternoon by the sun.

I took off the big Silver Grey and because I thought it suited the light and water clarity, put on another two sizes smaller for the tail and a Dusty Miller, also an eight, on the dropper. Salmon in this area average less than nine pounds and casts as light as will carry the flies are stout enough – mine was a 8/5 or in nylon terms about eight pounds B.S.

Before us, when Angus stopped the boat's forward way, was a long reach of stony shore – stretching to the left into the bay and right out to the headland; we had ten feet of water under us and the Stone – so far hidden – lay somewhere in the hundred yards or so that separated us from the beach. The wave was a fraction smaller than I like but the Dusty Miller, tripping and arrowing the wave, twinkled in the sun; it looked right. In spite of its height the loch water was very clear and I could see the Silver Grey on the tail in an occasional wave, travelling smoothly like a tiny grey ghost. I felt it had a beckoning look but then I do not quite see things as a salmon does in spite of forty years of hard trying!

One fish at least seemed to like the fly, appearing momentarily in the hollow of a wave as he turned down without apparently opening his mouth; we pulled off and were over him again in two or three minutes expecting to raise and hook him on the Dusty which worked beautifully in the sunlit broken water – he did not come again. 'We'll get one by the Stone,' Angus said. Overside the bottom showed clean stones of all sorts and sizes without trace of silt or algae and twenty yards ahead the huge golden blur that was the Yellow Stone we had come so far to seek. On this day of extraordinarily clear water and light it showed as an enormous flat-topped boulder, a hundred tons perhaps of water-worn gneiss, one corner just awash in the wave hollows. Round it were others, smaller and lying deeper – shadowy humps separated by clefts of darkness – round this loch at least there is ample cover.

I fished the Yellow Stone and all around it with such skill, care and concentration as I can muster and if I know anything at all, I do know this game, this kind of water and these conditions – and I could not move a single fish, much less provoke a taking rise. Although these salmon will not stand much fishing and the first time over is by far the best chance, we persisted for a while – quite fruitlessly. Angus pulled out and away saying things in Gaelic about the Stone while I said them in English. Fresh fish were here but if they were not in the best-reputed lie, where were they?

Along the beach on our left was a small round inlet – a bay within the bay so to speak and fairly in its centre was a pile of rocks the topmost two feet above the waves and others about it awash. We were drifting without plan now and I kept looking at the rocks. Angus knew what I was thinking and said, 'It's too shallow in there – full of rocks – and if we get in there we'll have trouble getting out – but we'll try if you want.' I wanted.

We bumped twice on the way in but Angus twisted the boat head to wind and wave and I fished island style off the stern. The face of the rock-pile was fished with the tail fly just about touching the stones and this was about as near as we could go. At one moment the stern post would hit a boulder just under the surface and the next there would be three clear feet under the keel. I threw a longer line and fished the right flank of the rocks, then as the boat sidled to the left, the other side. And here among the breaking waves came the good solid pull without sign on the surface. Angus said, as he always does when a fish is hooked, 'That's him.' That fish had us in a bad place and fortunately his move when it came was out to sea and not shorewards where it was impossible to follow. One big advantage of the Hebridean way of fishing by backing the boat down on any lie of limited extent, is that if a fast runner is hooked and goes out, the boat is in the best possible attitude, bow to wind, to follow. We pulled straight out after our fish now. He did one threshing double somersault in a turmoil of broken water thirty yards away but we soon had him in the boat – fifteen pounds and fresh enough for his flanks to carry the red veinings of lost sea-lice.

Angus made a rather rude gesture in the direction of the Yellow Stone as we passed outside it and started the motor; I remembered then that the lunch was still in the bag. We ate it as we swooped and rolled south through the islands with the motor at half-speed. Like most fishing days it *might* have been a big day but I was content enough to look at two fish in the boat – neither from the famous Yellow Stone.

XXIV

Missing – A Run of Seatrout

IT HAS always seemed natural and understandable to me that the known and characteristic instability of animal and bird populations must be equally applicable to fish in general and to salmon and seatrout in particular.

True, our judgment of migratory fish populations can only be based on the evidence available to us which is, in general terms, the size of the runs. We can have no basis for estimation of deep sea stocks. All we can know of salmon and seatrout salt water populations is derived from observation of estuary or river mouth stocks, which will be feeding stocks of seatrout mainly finnock and small adults, and of salmon due to run and collecting in wait for running water.

In some years a run will fail to appear and anglers throughout that river system will speculate as to why. Sometimes, very occasionally, it is possible to pin-point an answer – a bad spawning season – shortage of spawners – an untimely flood plus frost or the like the appropriate number of seasons previously in the case of salmon – or perhaps sheer lack of water. In the latter case, the fish may have fallen victim to estuary nets – or they may run late – too late for angling or observation.

In the case of seatrout, spawning usually every year after maturity – the reason for a run failure is infinitely harder to identify because a seatrout run will, if we exclude the early-running large fish, consist of mixed classes – finnock and small and medium class adults. These fish are, of course, the product of a number of spawning seasons and almost all, including a small percentage of finnock, will themselves spawn. It is admittedly true that generally the peak of the finnock run tends to occur after the peak of the adult run but once the summer run of spawners gets under way there will always be a large number of immature fish mixed in it.

Poor seatrout runs are known of course, and there are few river systems where fluctuation in numbers is not evident in the records of the years. If we accept that instability of population in wild life is related for the most part to instability of food supply and hence of its character then we can argue with some reason if not with complete conviction, that a 'failure' of food supply in the usual haunts of feeding and growing seatrout will delay maturity in maiden fish and 'ripening' in older seatrout. A direct result could be failure to turn up at the appropriate season at the river mouths. Whether seatrout failing to find normal feeding conditions inside their normal range of feeding movement go far afield in search of more plentiful feed, I do not know, but I am pretty confident that they do and that this often underlies their late – perhaps after fishing season – arrival in their rivers. It is a proven fact that big seatrout do travel far and I think this is why they are so often mixed with spring salmon.

What then of the short-absence fish – the finnock, herling, whitling, sprods or school peal? These fish have been in salt water for a few months only and in that time have assuredly not been far from the river of their birth. Indeed we know from study of their habits that the smaller seatrout whether immature or mature, are mainly estuary feeders, though some may wander quite far along adjacent open sea coasts. One can hardly argue that such fish will find themselves far from the river at their due running time. Even with a food shortage to drive them away I feel that the proximity of the river is too strong an attraction for them. Thus, with these considerations weighed, we are driven to the conclusion that whilst rich feeding may draw the larger, older classes of fish to more distant waters and result in non-runs or very late ones – we must look elsewhere for the reason behind the non-appearance at due time of finnock and young classes of seatrout.

Some observations in the season of 1968 have led me to the belief that a quite different factor may account for the absence of small fish – the 'condition' of the water supply in the river and its effect on the estuary where the fish feed in spring and early summer and collect prior to running in July and August. Water 'condition' in this context means its temperature and oxygenation.

We are all familiar with low water conditions in the ordinary sense and agreed on the desirability of a spate to aid the running fish but after some seatrout study (I caught my first seatrout in the summer of 1910) I know that determined seatrout will run in very low water and at night will struggle through the thinnest of stickles to populate a river unless stopped by some absolutely impassable barrier. Fish up to, say, two pounds will wriggle upstream in incredibly thin water and I have many times watched them doing so. In 1968 I watched and waited for them to do so in two rivers in the Outer Hebrides – and they did not!

A few came in and the catch in the peak months of July, August and September was one half of average but the big runs normal for these two rivers never materialized. The season was extraordinarily dry for the normally wet islands and at one stage salmon were dying in numbers in the small bare rocky pools of one river where they lay in blazing sun and stale, very warm water. Most of these fish, brought in by a small quickly spent fresh, already exhibited the white spot which develops in bright sun and high temperature in shallow sandy estuaries if the fish are held up by lack of water in which to run. A few seatrout were there but of course would find shade more easily than the bulkier salmon and apparently suffered less.

But there were not, for all practical purposes, any finnock or small fish in rivers where normally the pools are paved with them and where eagerly-taking finnock are a nuisance. One sea-pool, where normally hundreds of ten- to twelve-ounce fish come in with every tide, carried a few scattered individuals and the run had not arrived when I left the scene in mid-September. At no time was there so little water that these fish could not wriggle up into the sea-pool at the top of the tide.

Nor is there any evidence that the normal shoals of small seatrout were *absent* from the estuary prior to the establishment of drought conditions though, on the other hand, there is no evidence that they *were* there. The simple fact is that the estuary usually teems with seatrout and their presence is something that the watchers take for granted. The abnormality of the season only became noticeable when the rivers fell lower and lower, the sun blazed day after day and those seatrout which had penetrated the lower portions of the rivers actually went back to sea!

I know something about 'false' runs of seatrout which occur at various seasons and from which many return to sea after quite considerable penetration into the river systems. But the fish which comprise the main run in all rivers are true runners which do not return to salt water en masse when behaving normally.

In 1968 in the Hebridean river I speak of, this happened. Those fish struggled into the river and found the low, stale, warm water so repugnant to them or so inadequate for their needs as to cancel out their driving urge to press on. They turned tail and went back to sea. Thereafter the finnock, numerically the strongest class, were there in small numbers, the small adults from one to one-and-a-half pounds were totally absent and the average weight of the several hundred fish caught on the fly went up to about two pounds though, as already noted, the total catch in numbers fell to half the average.

The area concerned is one of heavy rainfall and sustained hot sunshine and consequent droughts are very rare. Local races of both salmon and trout do not meet, and I suggest are not adapted to meet the conditions of the summer of 1968. Fish native to southern rivers, for example, might be so adapted – their parr years might give them greater tolerance of low, warm water and bright light.

The weather certainly affected such island fish as did run in strange ways. I have already mentioned the white spot on the salmon. These eventually, on the head of the fish in fresh water, become ragged areas of raw, red tissue and I expected to see some secondary infection since in some pools the fish were crowded together, but no fungus was seen on any fish which came to hand.

Both salmon and seatrout did, however, behave untypically from a fishing point of view. Both were incredibly dour and both had suffered, in my opinion, in many ways from long exposure to hot bright sun in the shallow estuaries. They were languid and when they did take the fly – did so at depth. It was rare to see a brisk, fresh rise and this was especially noticeable with the seatrout. A surface fly was almost useless and such fish as came to the net, took the fly only when fished very deeply and slowly. Another feature of a strange season was the willingness of otherwise stiff salmon to take a deep slow fly in dead calm conditions and ignoring it in good fishing wave and light.

The mass return to the sea of seatrout was noted by other observers, men living all their lives by the rivers concerned so that the movement is no fancy of mine. None of us had ever seen it before nor imagined conditions totally intolerable to the fish. As to bad taking, there is a tradition that both salmon and seatrout, held up in the sea and unable to run at their due time or thereabouts, are bad risers and never take. I believe in it.

XXV

What Makes a Good Pool

THESE notes could equally aptly have been titled 'What makes a bad pool' because their purpose is to compare one class of river pool with another and to consider what may be the reasons for the contrast and that which exists between one holding-ground and another, apparently similar, in a loch.

River pools can be improved by the provision of lies, by minor diversions of current – though this can be risky, and by a variety of means which are well-known. The creation of pools where none existed is a vastly different affair and more often than not the only result is a well-justified conviction in the mind of the angler that there are certain things which salmon and seatrout cannot be forced to do! For example, they cannot be made even by all the understanding and cunning of fishermen and the skill of watercourse engineers to like and lie in a place not of their own choice. Vast amounts of money have been wasted in trying to guide salmon where, clearly, they did not want to go and in the endeavour to create new paths for seatrout and runs of them where none existed. The failure is not quite absolute but the history of fisheries generally shows that the more drastic the man-made change the less successful it is especially with seatrout.

The outflow from a loch can be improved to provide a good lead-in for entering fish – spawning ground in feeder streams can be opened up and obstructions removed. Something can be done in water level control combined with sluicing of dams to yield artificial spates, but a salmon fishing loch cannot be created as a river pool can. Comparisons of good and bad fishing grounds reflecting their status as holding places in lochs can thus only refer to *natural* physical characteristics. To be sure, there are good and bad places in rivers which are also as nature designed them and as she continues to work on them through the medium of flowing water, shifting gravel and the rest.

Some unproductive natural pools have always had that reputation, some acquire it following, perhaps, a phenomenal flood which violently alters the river bed. Most of us have seen good pools go off for what turns out to be this reason when later the bed of the river can be examined in very low water. This is simple and inside human understanding: what is *not* comprehensible is why one pool is consistently productive, whilst another with similar attributes of flow, cover, depth and apparently attractive lies, not only never yields a fish but is never seen to hold a fish. I know pools, beautiful looking pools in which no one has ever seen a salmon. Some are shunned by the seatrout also which is significant because it reinforces the suspicion that there is no comfort in them – perhaps because of the behaviour of undersurface current. Two of this kind are close together on a Scottish river I know well. They look just right and every stranger will fish them with great care and proportionate hope: they are not condemned as useless for lack of fishing. Fish pass through them but do not stay. There are deeps and shallows, rock ledges and flat slabs, big stones and gravelly margins – and no fish. The strange thing and a fifteen-year-long puzzle to me, is that they lie at the upstream end of quarter-of-a-mile of very thin broken water. Seatrout swim through quite easily but salmon have to struggle through all this rapid water. Why do they not rest in these pools which in appearance are ideal in character and ideally sited as resting places?

Lower down that river there has been for many years a sea-pool – so called. It is not a sea-pool in the usual sense of the word because the tide does not come into it: it is simply the lowest pool in the river. Below it is several hundred yards of broken water – backed up a little by a very high tide but usually quite fresh. The 'sea-pool' was contained for many years by a low dam provided with the usual gap and consisted of a long shallow flat in which the only cover for salmon consisted of undercut peat banks. Clearly it was too shallow at about two feet though it did hold fish. The dam and water level were raised by about a foot with a considerable improvement in the pool's fishability because the water now caught much more wind. The pool lies practically at sea level in a deep glen and a day with enough wave used to be an event. It is difficult even with the higher water level to get a big wave though a brisk wind against the slow current will create a chop in which it is a joy to work the fly. Fish with sea-lice are always present and it is rated a good pool.

Below it and filled directly by the flow over the dam is a new pool made when the dam of the older one was raised in level. This too is retained by a boulder and concrete dam with excellent gap and looks if anything a better pool than the older one. It has greater depth – six or eight feet, a steady and strong current, deeply undercut banks, clean hard bottom with scattered boulders and a very attractive-looking tail. An odd seatrout comes from it – never a salmon. It is fished repeatedly and is constantly under the eyes of fishermen and gillies; fish will be showing all over the upper pool but the lower one is dead always. Surely an odd fish must stop there? It is not a question, in my mind, of it being too near the sea and of fish always pushing on upriver because they do stop – some all season, twenty yards higher upstream. So here is an example of a natural watercourse converted at this point into an extremely attractive-looking pool which is a failure. Salmon do not like it but they like the less attractive (?) pool higher up. Very odd. This is where natural divination goes astray – where the most dedicated student of salmon rivers will make a mistake. I have always looked askance at this lower pool because I knew the river before it was made and that its construction was a gamble because fish might not like it. The doubt turned out to be justified but if I had been a stranger, seeing the new pool for the first time, my natural divination would have classed it as the better pool of the two.

To find an equally perplexing situation in a very large body of water, I go south, still on the gale-swept Atlantic fringe, to a lough in the north-west of Ireland. No question of human interference arises here, my comparison is between one area of lough floor which always holds fish and another which does not. Both locations are as they were left at the big melting, by the fourth and final age of ice about twelve thousand years ago. Grit-laden wave action, silting and settlement have doubtless wrought on the ancient rock fragments torn from the mountainsides by the valley glacier that littered the lake floor with boulders, but otherwise the boulders are as the ice dropped them. In fisherman's terms, we have two large areas filled with boulders of all shapes and sizes, some just awash, some showing only in a big sea, some two, three or four feet under the surface. Dark caverns gape between them and provide cover for salmon, the rocks themselves clean resting places. Both boulder areas are on the west shore.

The one producing salmon lies as a band of irregular width some fifty yards offshore and with minor breaks runs two hundred yards long and fifty wide and gives an average depth of about four feet. Except when the lake is high a few very big boulders show mossy humped backs above the waves but fishing the area one can see submerged round stones of all sizes in every direction. This is the kind of ground that salmon love and it fishes well, because even a big wave raises no mud and the stony fastnesses are full of shifting lights and shadows as the marching waves roll overhead. Such places flatter the fly: in the wave it works well and in the ever-changing ground colour it looks well. The water is clear and every movement of a fish coming to the fly can be seen as can every twist and turn of a hooked fish as it is worked up to the boat. Even now with so many years of salmon fishing in loughs behind me, I can never quite decide which is the supreme moment in the fishing – the take when the line draws tight and you're 'in him' or when you first see your hooked fish clearly, shoulders and back awash as he surfaces fifteen yards away or bright and subtly shot with colour in the wall of a wave.

Fish hooked on this boulder-strewn shallow tend to play high in the water: they make sustained fast runs very often. They do not have the depth to go down and give the rather dull play that a deep playing fish often provides. We do, of course, take them out and away from the fishing area to play them but even there the water is not more than ten feet deep. I have had three fish in a day from this location and ones and twos are the rule providing wind and weather are reasonable.

Half-a-mile away along the shore and separated from the beach by a fifty-yard strip of moderately deep water, is a low wooded island its foundation a huge heap of the boulders typical of the lough. In Chapter XX I described a crannog. This island, however, is natural though ancient man has left traces of occupation. Today it is a microcosm of nature on the lake; the taller trees carry a small heronry and the pigeons nest there. Small birds haunt the undergrowth and in the spring the mallard nest in a carpet of bluebells and wood anemones. One of the most beautiful things to be seen is the soft-eyed duck sitting calmly – pencilled brown among the flowers. Small trout feed, dart and play in the shallow margins under the trees and in deep water nearby the char dimple the surface on a summer evening. To me these things are not distractions from fishing, they are a part of it. Too many anglers lessen their pleasure by being heedless of them.

At the outer end, the seaward end so to speak, of the island, the boulders are thick in the shallow water which, however, gradually deepens as one moves out from shore until the boat floats over another stony jungle precisely similar to that which I have already described and which salmon frequent. For a hundred yards out from the island and to either side of the main mass, there are two, perhaps three, acres of water four to five feet deep, clean and stony and filled with golden-yellow boulders. I do not claim to know holding places when I see them – I only know at sight places which by all standards of experience *ought* to be holding places. And this is one of them. I have never seen better-looking water but in all the lough's fishing history no fish has ever come from it. To all appearances it is even better ground than that half-a-mile away which yields fish throughout the season but, again, salmon do not like it.

Salmon entering a lake in the spring commonly follow the shores and may exhibit a decided preference for one shore. Some fish settle en route, others go on to proceed up another river and the majority of those staying in the lake till autumn spread widely as the weeks go by and may be found anywhere on suitable ground. Now I can understand that the first run of spring fish in my lough, questing along the shore, will travel inside the island rather than outside and will on this line come to rest on the good ground which is directly in their path – missing the apparently equally good ground further offshore outside the island. But I cannot understand why the subsequent scattering of the fish all over this part of the lough does not populate this second bouldery shallow. And, as is true of some other places in both river and lake that I know, the seatrout coming in hordes in July also avoid it. Good and bad grounds have similar exposures and get the same prevailing wind and much the same wave strength and drive; it is probably slightly rougher in a given wind-strength on the island beat but that cannot alone account for its being untenanted by a breed of salmon which exhibit every sign of liking rough water.

I often drift the island beat but I have never seen a fish there, much less risen one. I can drift it all day, if I have the patience, in the certainty that no nearby plunger will startle me out of reverie. The fish simply are not there.

There are, in every salmon loch wherever it may be, as in every river, recognized lies and when I said above that fish will spread widely and may be found anywhere I meant in such lies wherever they may occur. A lie, as distinct from an area of holding ground, may be very small – as isolated rock, the end of a wall running out into the lake, or a small hollow in the delta off an entering brook. It may, often does, hold one fish at a time; if that fish is caught another will move in but quite often the lie may be untenanted – it depends on the density of the stock. If then such little holes and corners, lost in the immensity of water, can attract fish – how extraordinary that a large area abounding in lies should be unoccupied. Season by season I continue to puzzle over it.

It would be cause for shame if in a lifetime's fishing I had not come to know what in general terms constitutes a holding place for salmon in lakes; rivers are much simpler to read. The critical words here are 'in general terms' – in absolute and precise terms I quite evidently do not know which is properly humbling and a very good thing. One begins to know about fish and fishing with the realization of how little one really knows.

XXVI

Irish Interlude

ONE of the infinite joys of fishing resides, for my taste, in the extraordinary range of characters one meets, more or less picturesque in word, deed or appearance. Millionaires whose fishing garb can be valued in shillings and poor men 'correctly' attired for the river, men with dilapidated gear who catch fish and beautifully equipped others who do not. However it shows itself, the personality of the man usually emerges in the boat, by the riverside, where everyone eats, drinks and sleeps, or wherever the atmosphere is all fish and fishing. There is a special enjoyment in real whole-hearted fishing which comes from mixing with other anglers, gillies and countryside characters whose ways are all on or about the waters. Every so often one meets a real 'character' though I confess I seem to see fewer of them than was once the case; perhaps I am not as perceptive.

Years ago in Ireland, at the spring fishing particularly, 'characters' abounded and in those few weeks life on or about river and lough could be a riot of eccentricities and a sheer delight if you had an eye for personality and a suitable sense of humour. The utter inconsequence of minor happenings in the more remote parts of the country and the beautifully quiet and slow tempo of life generally, set the scene of course.

Our circle was small and made up of certain village worthies and of visitors of such long standing and so well known as to be accepted almost as local inhabitants. There was much keen if under-cover rivalry among we fishers on the lough and the small village hotel was ever in a buzz of salmon talk, some of it coming rather ponderously from two elderly legal gentlemen of serious mien and of great repute in Dublin. Inseparable, they fished together daily, the large Mr. Campbell in a stern well down with poundage and authority, and the small Mr. Gorman in the airy bow. They employed a gillie from some miles away and as a result we heard little of the goings-on in that boat but they did bring in occasional fish.

Mr. Campbell, as befitting his age, rank and seniority among us, presided each evening at the head of the dinner table, having previously applied himself to the contents of a special bottle of the crayther which he and his partner favoured and of which our inn proprietor once said, 'Shure you wouldn't be usin' ut to take the paint off a cart.' Mr. Campbell had an enduring complaint about a mouse which haunted him and the fact that he was forced to sleep clutching a walking stick. This special mouse, among the hundreds that infested the place, was wont to comb his whiskers and perform his general toilet in the small hours seated comfortably on Mr. Campbell's chest – underterred by the heavings, or the whisky-laden snorts or by nightly swipes of the walking stick. This by the way. Some of the conversation that adorned those dinner sessions ought to be preserved but this is a book about fishing and I must confine myself to one example and incident.

It occurred one night when the house dog – an

unkempt cocker spaniel – seldom seen 'above stairs' invaded the dining-room and seated himself abreast of Mr. Campbell's chair, gazing fixedly at that gentleman and refusing to 'get along wid yez'. I wondered momentarily whether under the pompous bulk there lay the kind heart of an animal lover and the dog knew it. Not a bit of it – furious ringing of the bell presently brought an agitated maid. 'Nora,' growled our chairman, 'take that dam' dog away. I think he's out of his mind. He just sits and stares at me and I'm beginning to wonder if I have a right to me own dinner!' 'Ah well sorr,' said Nora, 'tis aisy to ondherstand in a way. Ye see we had a big upset below in the kitchen wid the crockery all broke an' Nipper here knows ye have 'is plate!'

Out on the lake we often had Father Dermot the village priest, a remarkable character of ever-changing enthusiasms – at that time concentrated on fishing. He had sold his collection of photographic gear and accumulated a house full of clocks which he had sold to buy an ancient and very bruised car presently traded for a boat (oddly enough a good boat) and fishing gear. Every free hour found him fishing alone and I only ever saw him hook one fish. I was lunching in the lee of a hillside wall with my gillie John one day and Father Dermot was fishing in the bay below us, sitting as usual in the stern so that the wind caught the cocked-up bow and swung the boat all ways. It was so foolish that our amusement turned to irritation and I ceased to watch until roused by John saying, 'Begob, Father Dermot's in a fish!' He was. The breeze was light and the fish very helpfully stayed reasonably near the boat, which with the fisherman so far aft simply rotated on its stern and turned with the fish. Father Dermot held on and as the fish circled so he rotated on *his* axis – long black coat-tails in the water at the stern. Somehow in the turmoil his broad-brimmed black hat went overboard and bobbed away downwind. Then everything went fairly quiet. We could see the bent rod but no sign of a splashing fish though there was at one stage some scrambling forward in evident search for the gaff. There was a certain amount of standing up and sitting down again and of coat-tail dipping and after half-an-hour we concluded we had better see whether we could help. Father Dermot, after a few weeks' fishing, knew it all and was apt to treat advice as a reflection on his skill and vast experience. We pushed off and I took the oars while John prepared to transfer to the priest's boat. 'Can John gaff him for you, Father?' I asked. 'He can,' came the answer, 'if I can get the divvle up.' The fish was somewhere under the rod-tip and that was almost in the water: we backed up on the opposite side and John climbed in while I pulled quickly away. 'Ye'd better get him from under the boat, Father,' I heard John say and the complete angler's reply, 'Easier said than done, me boy, but he'll come to no harm. Sure, he's been under there this last ten minutes.'

The end of that episode was the gaffing of a very tired twenty-pounder. And the last time I saw Father Dermot he walked into a friend's house where, in the larder, the friend and I were packing a salmon in rushes – incidentally the finest packing of all to keep a fish in condition if the rushes are thick, green and long enough. 'How is your anti-poteen campaign going?' I asked the priest who had, I knew, been conducting a drive against the illicit distilling of spirit by the wild boys of the neighbourhood. 'Fine,' said he, 'ye won't find a drop in the parish.' A two-gallon jar of the stuff sat on a shelf within a foot of his nose while he spoke and the situation appealed to me greatly: my friend said later that for his part it had him 'destroyed entirely'. Apart from anything else, the police-station was next door.

Fishing was the main topic of conversation in *that* establishment too for the sergeant in charge, a quiet man from Connemara, was a keen and highly proficient trout fisher. Between us we took some bags of trout of high average weight from the small river behind his office – he by orthodox wet fly and I on the drifting nymph fished downstream on a loose line. Before I knew our sergeant well, his early-morning perambulations used to puzzle me. Sometimes he quite clearly put in an hour on the river, but usually he passed my window on the stroke of seven-thirty, heading smartly up the village street. In five minutes he would be back at half the pace and always with hands clasped and wrists crossed behind him. One morning I put my head out after he had just passed and beheld in each clasped hand – an egg. He became famous after an incident one Easter-time. There had been the spring races and the village in the evening was full of wild men with money to spend or losses to drown and the three 'hotels' had ranged barrels of porter round their yards – the bars would not hold a fraction of the crowd – and there was drinking and pleasant fighting and lots of fun. With a friend I was paying a social call on the sergeant when a panting small boy arrived to say 'they were after killin' one another in Dooley's yard'. The quiet Connemara one broke off the tale he was telling of a salmon in the main river, put on his uniform cap and departed. He was back in a very few minutes, put his cap carefully on its shelf and with a little hand-dusting gesture said, 'Now phwere was Oi at?'

In those few minutes he had quelled a rather serious fight without his two guards – busy elsewhere – by wading in and speedily cracking the contestants heads together to knock the fight out of them.

Fifty miles away from that village and its adjoining and very lovely lough is another lough which is not only closely restricted private fishing but has that blessed feature, a run of salmon from April onwards. Not vast numbers of fish but enough to provide a constant chance of a fresh fish. A good run of seatrout comes at June-end – water permitting. I spend as much time as possible on that lough, aided sometimes when my faithful James cannot attend me, by one Thomas. Thomas the strong man, the greatest puller of boats in that county of boat-pullers – Thomas the breaker of oars is a great 'character' with a flair for colourful exaggeration. He once reduced a friend of mine to uncontrollable sniggers by telling him that whilst on good days the salmon would be jostling the boat in their eagerness to get aboard, on a real bad one fishing was as futile as 'T'rowin' peas at the Gineral Post Offis wid the object of demolishin' ut'.

Last spring on a bright windy nor'-west kind of day, I was looking for a salmon on a gravelly shallow near the lough's outlet, with Thomas at the oars. Here the lough floor rises gradually while the shores narrow to the place where the outflowing river begins to exert its pull. Where river ends and lough begins is very thin water flowing swiftly under the arches of an ancient multi-piered bridge and there is, moreover, a long shallow of broken water extending downstream from the bridge. Incoming fish, having struggled through the broken water, quite often settle temporarily just inside the lough before moving into recognized lies up and around the shores. On the day in question I was searching for these fish in a series of short drifts outside the shallow water. I had been having minor trouble with Thomas who wanted me to fish other water about a mile away – the passage thence, of course, giving him opportunity to open his shoulders in the kind of sustained pulling he enjoys. Whether that morning I had more green in my eye than usual, or whether Thomas merely reacted in his own way to my obduracy, I do not know but a simple question from me produced a tale I shall long remember. 'Have you ever seen a really big fish in the lough?' I asked him (our fish average ten pounds and a big one is seventeen or eighteen). Thomas reflected for a moment and then nodded with his chin towards the bridge. 'D'ye see the t'ird arch forninst ye there?' I saw it. 'Well, wan day last winter – December it would be, I was standin' on the bridge and the fish spawnin' on the gravel there – dozens of thim. There was wan in among thim – the biggest salmon iver I saw. T'was a hin fish wid a back as wide as young Kathleen's she-ass. Ye couldn't put a pin between the fish workin' at cuttin' an they was that through other that ye couldn't count thim. The big wan keeps goin' round and round lookin' for a space till she's caught wid her side to the current an she's swep' away an there she is ondher me feet – jammed across the t'ird arch there wid the wather pilin' up and the lough beginnin' to rise. T'was the biggest fish ever I saw.'

XXVII

Kelts to the Healing Sea

SOMETIMES in a bad spring with few fresh fish about my irritation at the eager taking of a kelt has been merged with a faint gratitude. At least his taking has allowed me, after the inaction of a long winter, to feel the rod bending to the pull of something vibrant and alive. And until his wriggly pulling conveys the 'feel' of a kelt, there is the joy of a take – the stoppage of the fly that can so easily mean, in due time, and if the off-season has not robbed you of your 'hands', a lovely clean fish on the grass, tan and mauve with the sheen of a fresh one and the smell of the sea on him.

I do not mind the kelts. They have made their contribution to their posterity and our river stock and however frayed of fin and tail or blotched with fungus, however shadowy their relationship to former elegance – they have the dignity of certain wild creatures and a pathetic one of their own. I am aware of the laughter of the cynical and the hard-hearted but I do not share it. I land my kelts in the net, treat them gently and put them back on an even keel with gills working evenly and am happy to see them swim actively away.

I know that the proportion of second spawners is low and that saving all the kelts will not significantly alter the size of subsequent runs but I find this period in the life of the salmon kind not only an interesting but a necessary study. It is not enough to accept simply that after spawning the survivors drop back to sea. There are left weeks, maybe months, of struggle to live, of deathly quiescent periods in the cold depths of the big pools, of weary strivings in buffeting rapids, of drifting down the flats and of the constantly sharpening urge to feed.

That kelts do feed in the river en route to the salt water is not universally believed and the majority indeed may not. It is often argued that if kelts fed in earnest in the upper reaches of river systems which are nurseries of young stock, they would speedily decimate the young of their own kind as well as those of the native trout. I think this contention must be weak until and unless we know whether sufficient kelts survive to become such a menace. If there were total survival of a spawning run and they all fed before returning to salt water, it would be a very different matter.

That the suspected small percentage surviving do feed seems to me to be beyond doubt. It seems strange if they should not since they have urgent need to replace energy, wasted tissue and general condition and have no reserve, as fresh fish have, to sustain them over protracted fasting – anything from two to four months. One considers too, the avidity with which they will take fly or bait – an apparent eagerness to take anything moving near them. A few months ago, these very fish were refusing to look at anything! That they take does not prove that they are capable of digestion, of course, and herein is the rub.

Whether food taken by kelts still in the river reaches the stomach and is digested I do not know but I do know that some kelts do feed because I have seen them doing so on a number of occasions – clearly and unmistakably feeding. One shallow pool I know well on an Irish river in April is always full of kelts. The pool bisects a long reach of broken water and its easy current obviously provides some rest for descending fish. Parr, small trout and early-going smolts are here too and I have many times watched the tenanting kelts in pursuit of the small fry. Good human eyesight is not quite that of a falcon however and I cannot therefore say I have actually seen the small stuff disappear in the maw of a kelt. But the activity and its purpose are beyond any shadow of doubt. In loughs I have thrice seen the water break and seethe with fleeing smolts precisely as will happen when pike are feeding among shoals of small stuff or when a really big trout is hunting the shallows. But a pound trout is a big one in these loughs and there are no pike.

Guanin is a waste-product of the body processes of the fish, derived from the blood and forming on the scales the irridocytes which reflect light and give the fish its brightness. It is difficult to see by what process kelts – whether of salmon or seatrout, recover the brightness which is an index of improving condition, without an intake of food.

In the biology of the salmon no explanation is evident as to why so many male fish die as kelts as compared with females. From my own observation I would have concluded that slightly less energy was consumed by the cock fish in milt emission than is used by the hen in extruding ova. In addition to the act of extrusion she has had the labour of cutting the beds. If we admit that the cock consumes a great deal of energy in driving off other males – including small cock trout and precocious parr, he should still emerge from the total process the stronger of the pair. But he does not.

Why, again, are some kelts so late in going seaward? Even if they are phenomenally late on the spawning grounds, it is hard to understand why they will linger in loch or river, as happens, till May or June. I have found kelts high up in Hebridean river systems in June or July and seatrout kelts, desperately poor in condition, in August. No one will convince me that these fish – though such thin and very obvious kelts – had not been feeding. But why had they not gone? It is possible that fish spawning very high up small feeder-streams in high water may be imprisoned in little pools when water falls and their retreat to the sea then retarded, but dry weather conditions in upland country do not last very long in the early months, and there are plenty of spates of varying degree to take the fish away. I do not, in short, think that the progress of spent fish to sea is wholly a factor of water height – something else is at work to make them linger.

In some areas the salmon appear to have habits in some respects very like those of the seatrout and it is possible that some well-mended kelts taken in the lower and middle reaches of some Highland and island rivers, have made a short visit to salt water and returned temporarily to the river before departing for a longer period of feeding and growing, as so many seatrout do. I have caught a number of strong, fresh-looking but very thin fish in July and August which were undoubtedly kelts that had re-entered fresh water with the fresh summer fish. Scales, on examination, showed barely perceptible growth since spawning – but the growth was there and looked like the beginning of sea feeding. This 'false run' habit is commoner in seatrout, of course, but I can recall no reference to it in the literature of salmon life history and behaviour.

Speaking of behaviour, there are some interesting aspects of kelt downstream migration in the context of how they move. Where the contour and cover of the banks will permit one can sometimes, rather rarely, see the fish, and detect at once the prevailing head upstream attitude, except in very dead water which they will swim through before swinging round at the next stretch of broken water. Arrived at the point where the current's pull begins to be evident, however, fish heading down will turn head to stream and begin their reluctant descent through rough or shallow water. One watches with a curiously strong goodwill. I have seen small companies of kelts, fighting the draw of water over the lip of a weir yet gradually giving ground as if they knew that they had to let go, until the leader, always the largest fish and furthest downstream in the group, ceased to fight the stream and was swept over tail first. Then followed the others and finally the hangers-on, the seatrout. The whole operation might take quarter-of-an-hour, the fish hanging three feet above the lip, drifting to and fro in the clear gliding water, every nose upstream until, behind them, the big fish slid away.

I would like to know a great deal more about the lives of kelts in the river but to learn will take more years than I have left: it has taken me a long time to gain what I know now – so darkly hidden are the ways of the fish going back to the healing sea.

XXVIII

'Goodbye Now'

NOT farewell, but goodbye now, the parting word that is a timeless re-welcome and in the tone of it a benison and a wish for the days to come. They don't like parting with people to whom they say 'goodbye now' in the country where I hear it so often and it has its own little warmth of memory for me.

A fisherman, if his heart is in it, is a man happy in many blessings not least of them a retentive memory. The retained image is not always of great fishes brought to net, of this or that strategem, but of mountain, river and lake and their weather, of creeping and winged things, of water plants and grasses and trees and the honey smell of gorse and the crying of oystercatchers, of good companions and of places. Small things have great import as components of the whole that makes a fisherman's world. In them reside many of his greatest joys and afterwards they put colour to the canvas of recollection so that scenes in detail come at once to life again. Being part of the immediate background is part of the craft of fishing and of the process of catching fish but there is a wider background which belongs to places and atmosphere of places. You fish in the western isles and you fish in a world where the ocean is at your shoulder – you share the surge of wind and wave with the seals on the skerries, the soaring eagle and the shining salmon that carry the tang of salt far into the hills. You will always have the gold of the tangle at the back of your eyes, the machair of a million flowers and a network of tideways and lochs to lure you as it drew the Norsemen for other reasons 'west over sea' a thousand years before you. You probably catch a great many fish and some part of your fisherman's joy will come at the end of each good day. The rest is in what the Islands have given to the living day and the memory of it.

Fish in my beloved Donegal and in the west is the same wild ocean creaming below vaulted cliffs and the loveliest mountains in Christendom on one hand or the other as you go about your fishing. No eagles here but all else and for added benison such flame of gorse, such fragrance of bog-myrtle and such blue of distant hill as surround no other place of angling. The past, turbulent and gentle too, is close to you here with the legendary Thousand Champions of Aileach each in his leather cloak, and the prince-priest Columbcille, the dove of the churches and founder of Iona. Even if you have no taste for ancient Irish history the name of Columbcille is everywhere about you as you fish – by this lough in 521 his birth-place, by this stream the ruins of the abbey where he was taught, this lough bears his name and so does that glen. Here are lonely loughs and brawling rivers, salmon and seatrout and trout, heather and myrtle and the tossing plumes of bog-cotton on ice-moulded moorlands mile upon mile and over all the wild wind from the western sea. This country is steeped in myth and in early history and it bears the very air of it: you do not fish less effectively because the past is at your elbow and hereabouts, only yesterday.

Yes, the essence of places is truly in the air about them and all the life that is in them, to be felt and seen and remembered as you tread softly in your fishing.

A good day with the grilse is a thing to be cherished so rare is it, yet of my good days with the small salmon, I recall as immediately as with the fish of that tumbling little Irish river, the royal fern that here grows waist deep on the banks – such fern as I have never seen elsewhere. Osmunda grows small in my Devon garden; passing daily, I see it through the mist of spray from a roaring brown river and the lovely slim shapes of the grilse laid among the fronds.

Salmon fishing in Donegal is like no other to me. The fish are strong, rough fighters and in the lough hard runners after the hooking. In that country they fit their background bringing the restless life of the sea to an untamable landscape and bearing tradition more akin to the lawless pursuit of salmon in dark roaring waters, than to any refined sport of salmon fishing. Sunlit loughs and zephyrs of summer belie the fancy but your true Donegal days are the wild ones when the boat lurches in the rolling wave on a stormy lee, the great whooper swans, northbound, call overhead and the spring salmon new from the tide will – and you lucky – give you the run of your life. All of your perceptions are heightened by the very breath of this sort of fishing and you see everything in its designed place, the streaming lace of the mountain waterfall alike with the neat quillwort growing in the lapping waves on the lake shore, the scuttering take-off of tufted duck and the crow and herring gull squabbling over a long-dead lamb among the driftwood on the shore.

You could, if you were very sensitive, find pity for Columbcille far in Iona from this very land he so loved, those fourteen-hundred years ago. I think I know a little, a very little, of what must often have been in his mind – looking across the green rush of the sea to Ireland, and I too in my travels in a new and strident world, have been sick at heart for the crying of the gulls round an island in a lough of Donegal.

'Goodbye Now'

Index

Alexandra 39, 101

Black Doctor 28, 32, 34, 35, 69, 76
Black Dose 111
Black Fairy 33
Black Pennell 35
Blue and Silver 32, 101, 121, 156
Blue Elver 58, 63
Blue Zulu 32
Boat Equipment 171, 172
Brown Owl 36, 119, 121
Butcher 27, 32, 35, 76, 81, 82, 84, 101, 156, 190

Cast weights for seatrout 155, 156
Claret 23, 33, 109
Cloud cover and light 21
Colour of boats 46
Columba, Saint 106
Columbcille 244
Connemara 30
Connemara Black 32, 35, 156

Cormorants 199, 200
Corncrake Sedge 190
Crannogs 106, 187, 188

Devon river, a 90 et seq, 136 et seq
Direction of winds 20
Donegal 30, 106, 160, 244, 245, 246
Dunkeld 99
Dusty Miller 28, 32, 33, 99, 192, 212

Elver 64, 65

Fiery Brown 23, 33
Fly sizes 37

Golden Olive 34, 53, 156, 162
Greenland nets 196

Hebrides 34, 40, 79 et seq, 114

Invicta 81, 121
Iona 244, 246

Kerry 30
Killer whales 195

Lampreys 198
Lemon Grey 33
Lune tributary, a 122

Mallard and Claret 32, 37, 119, 121, 156, 190
Mallard and Yellow 32, 35

North Devon 124
North Devon rivers 198

Otters 195, 196, 199

Pennell tackle 101
Peter Ross 32, 121, 156
Porpoises 195
Prawns 62, 63

Queen of the Water 36, 119, 121

Ransome, Arthur 58
River Taw 61

Scandinavia 89 et seq
Sea food of salmon 60
Seals 195, 196, 199
Seatrout, habits of 112 et seq
Sheringham, H. T. 52
Silver Doctor 85, 99, 162
Silver Grey 32, 54, 211
Small rivers 41
Standard range of flies 31 et seq
Stewart tackle 101
'Supreme' spinning reel, the 149
Sweep 27, 34, 121

Taking times 18 et seq
Thunder and Lightning 32, 34, 70, 73, 109
Torrish 28, 32, 35
Trailed fly, use of the 212

Wales 30, 114
Water level and colour 22, 23
Watson's Fancy 81
West Country 30, 36, 114
Woodcock and Yellow 32, 121, 156

Zulu 32, 35, 156, 163, 166